Biological Pest Control through
ICHNEUMONIDS

The Author

Dr. Sathe Tukaram Vithalrao [M.Sc., Ph.D., Sangit Vishard, IBT (Seri.), F.I.S.E.C., F.S.E.Sc., F.S.L.Sc., F.I.C.C.B., F.S.S.I., FHAs] is presently working as Professor and Head, Department of Zoology, Shivaji University, Kolhapur. He has teaching experience of 29 years in Entomology at University PG department and 15 years in Agrochemicals and Pest Management. He has written 30 books and published 255 research papers in national and international journals of repute. He guided 20 Ph.D. students and completed 6 major research projects (from CSIR, DST, DBT and UGC). He visited Canada (1988), Japan (1988), Thailand (2002, 2004), Spain (2005), France (2005), South Korea (2006) and Nepal (2007) etc. for academic work. He is member of editorial board of eleven prestigious journals. He delivered 35 talks through All India Radio and international conferences and involved in Doordarshan, S.T.V. and B.T.V. programmes on useful and harmful insects. He published more than 35 popular articles in daily newspapers on insects and sericulture. He got several prestigious awards like "Environmentalists of the Year-2003", "Bharat Jyoti", "Jewel of India", "International Gold Star", "Eminent Citizen of India", "Education Acumen", "Best Educationist", "Eminent Scientist of the Year-2008", "Lifetime Education Achievement", "Lifetime Achievement in Zoology (Insect Taxonomy)-2009", Education Leadership-2011, Asia Pacific International Award-2012, Global Education Leadership Award-2013, etc. He is also working as Research and Recognition (RR) Committee member for Pune University, Pune; North Maharashtra University, Jalgaon; Shivaji University, Kolhapur and DBA Marathwada University, Aurangabad. He has been awarded several fellowships from different scientific and academic societies. He is Chairman of Maharashtra District Environmental Centre of NESA.

Biological Pest Control through
ICHNEUMONIDS

— *Author* —
T.V. Sathe
Professor & Head
Department of Zoology
Shivaji University
Kolhapur – 416 004, M.S.

2015
Daya Publishing House®
A Division of
Astral International Pvt. Ltd.
New Delhi – 110 002

© 2015 PUBLISHER
ISBN 978-93-5124-363-2 (International Edition)

Published by	:	**Daya Publishing House®**
		A Division of
		Astral International Pvt. Ltd.
		– ISO 9001:2008 Certified Company –
		4760-61/23, Ansari Road, Darya Ganj
		New Delhi-110 002
		Ph. 011-43549197, 23278134
		E-mail: info@astralint.com
		Website: www.astralint.com
Laser Typesetting	:	**Classic Computer Services**, Delhi - 110 035
Printed at	:	**Thomson Press India Limited**

PRINTED IN INDIA

— Dedicated to —
University Grants Commission
Bahadurshah Jaffar Marg
New Delhi – 110 002

Preface

Ichneumonids belongs to order Hymenoptera of class Insecta. They act as very good biocontrol agents. Hence, they have tremendous practical potential in Integrated Pest Management. Ichneumonids kill the pests either in larval or pupal form as a parasitoid. Thus, Ichneumonids are biopesticides scattered in the environment. The aspects like diversity, reproductive potential with respect to pest age, pest specificity and pest density; behaviours like mating, oviposition and emergence and over all intrinsic rates of natural increase will add great relevance in propagation and utilization of them in eco-friendly biological pest control. I feel that this book will be useful to farmers, students, teachers, scientists, industrialists and environmentalists. Author is very grateful to the Chairman and Secretary, UGC, New Delhi for providing financial assistance to the project 37/334/2009 (SR) under which the present investigations are carried out and Shivaji University, Kolhapur for providing facilities. Author is also thankful to Madhuri and Nishad Sathe for their help in many ways.

Prof. T.V. Sathe

Contents

1

General Introduction

The role of agriculture in the economic development of any country, rich or poor, is very crucial by the fact that, it is the primary sector of the economy which provides the basic ingredients necessary for the existence of mankind and also provides most of the raw material which fulfils the basic necessities of human and domestic animals.

Agriculture in India should plan to achieve at least 10 per cent growth rate. The need of production of food for the developing dynamic society is challenge to farmers and scientific community. Therefore, modern agriculture needs scientific basis and knowledge on the crops and their pests and diseases. Modern agriculture is one of the best options for economic development of the nation. Agriculture in India is the admixture of cultivation of variety of food and non-food crops. The major group of crops cultivated in India and Maharashtra (Figure 1) refers to cereals, pulses, fibres, fruits, oilseeds, vegetables and plantations etc. (Singh and Sandhu, 1986).

Cereal crops are the major part of Indian diet. They provide 350 kcal of energy per 100 grams to the body. Cereals contain 6 to 12 per cent of protein so it is also a significant source of protein, greater levels of B-

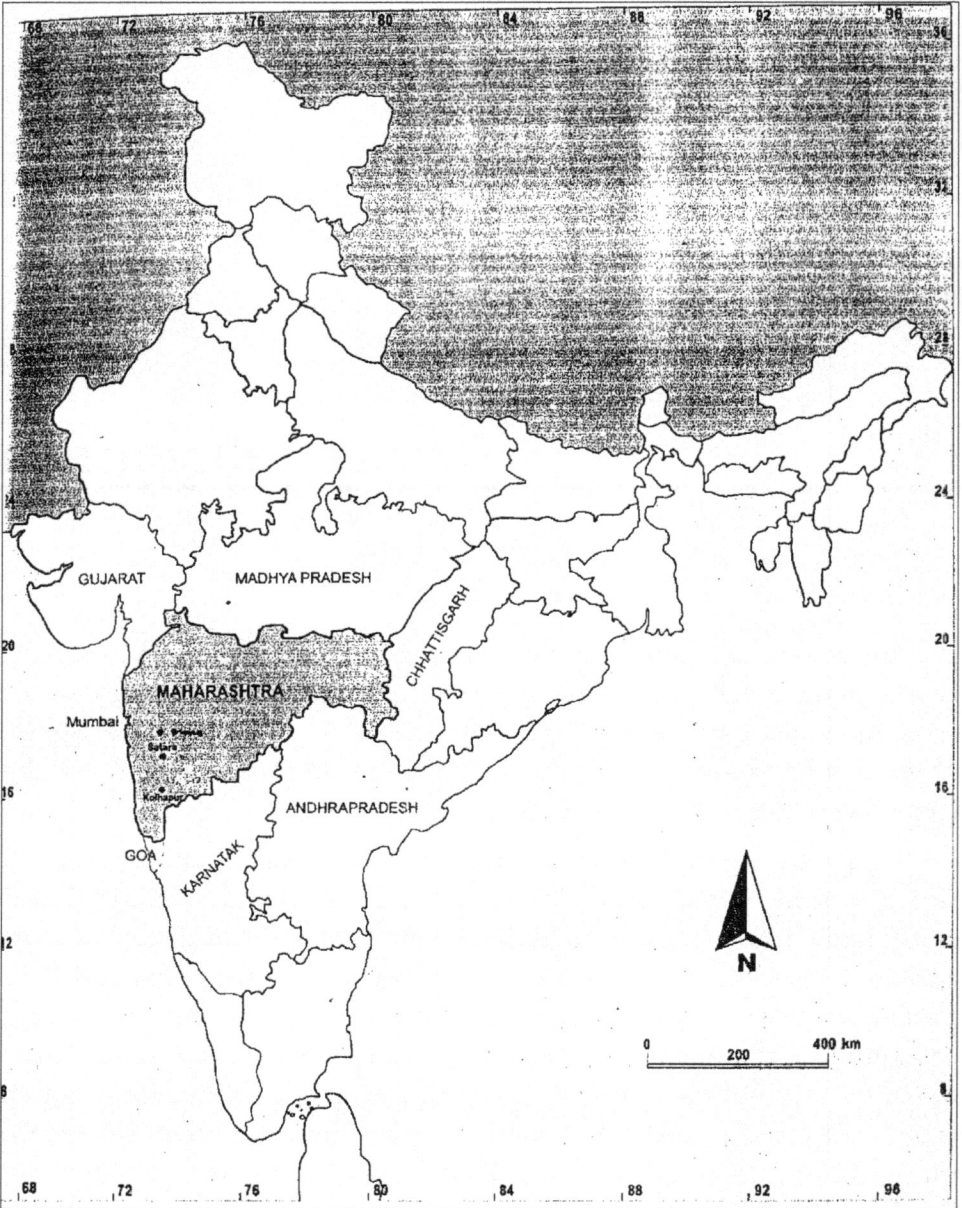

Figure 1: Map of India Showing Maharashtra and Study Area.

complex vitamins, dietary fibre and essential fatty acids. Cereals like millets also supplies minerals like iron and calcium.

Grain sorghum is the fifth most important cereal crop grown in the world. India shares about 12 per cent of the world sorghum production. Sorghum is one of the most drought tolerant cereal crops. Some sorghum varieties are rich in antioxidants and all sorghum varieties are gluten-free, an attractive alternative for wheat allergy sufferers (IACB, 2009).

As a cereal crop, rice ranks first in India followed by wheat and coarse grains. Rice continues to hold the key component for Indian diet. The demand for rice in India is 128 mt by the year 2012 and will require a yield level of 3.0 t ha 1 that is significantly greater than the present average yield of 1.93 t ha 1.

India with diverse soil and climate comprising several agro-ecological regions provides ample opportunity to grow a variety of agricultural crops including cash crops like sugarcane, cotton, grapes, etc. These crops form a significant part of total agricultural produce in the country. Fruits are rich source of vitamins, minerals, proteins, carbohydrates, etc. Hence, horticulture practices in India are given prime importance as nutritional security of the people. Thus, cultivation of horticultural crops plays a vital role in the prosperity of nation and is directly linked with the health and happiness of people. India produces more than 28.2 million tonnes of fruits is the second largest producer of fruits in the world, next only to Brazil and China. According to ICMR and NIN, Hyderabad, per capita consumption of fruits in India is only around 46 kg against a minimum of about 92 g as recommended. As like agriculture and horticulture, forest sector can also play an important role in uplifting of national economy and environment safety for the humans and other animals. Accordingly there should be 33 per cent forestation in each country for keeping environment eco-friendly.

Agricultural, horticultural and forest crops are attacked by several insect pests which damage stem, leaves, flowers, fruits, roots etc., by feeding upon them. Many soil inhibiting insects attack roots and kill the plant altogether inflicting serious injuries and destroying the root zone. On an average 35 per cent annual world crop losses are due to the pests (Cramer, 1967).

There is continuous fight between man and insects for choice of food, fiber and survival. Therefore, the protection of crops from various kinds of pests remains chronic problem. The introduction of high yielding varieties has also increased the pest problems. Many agrochemicals increase pressure on various crop ecosystems by creating pest problems. There are several options to pest control but, many became inactive and useless. Hormonal control, Radiational control, Genetic control, Behavioural control, Biological control, Pheromones, Allomones, Kairomonoes etc., are the important components of IPM. According to Sathe (2001) the Hormonal control, Genetic control, Radiational control, and sterilization have not at this time proved to be widely successful in pest management. According to Hebbalkar (1987) most of JHAs are thermo-labile and photo-labile hence, under field conditions their potential efficacy is adversely hampered. JH sensitive period as such is of few hours and under field conditions presence of mixed developmental stages of various pests is of common occurrence, due to which practical application of JHAs become more tedious and less effective.

Undoubtedly, chemical control provides quick results but leads to many serious problems in environment. The chemical control problems refer to:

1. Soil, air and water pollution.
2. Killing of beneficial insects like, parasitoids, predators, honey bees, etc.
3. Destruction of natural balance and ecological cycles.
4. Development of resistant in pests.
5. Pest resurgence.
6. Secondary pest-out-break.
7. Stimulation of the reproductive rate in certain pests.
8. Health hazards, etc.

The above problems suggest to search for an alternative to pesticidal use. Biological control is good alternative to the pesticidal control. Amongst the several options to pesticidal control *viz.*, cultural control, behavioural control, hormonal control, genetic control, mechanical, physical, anticidal etc., biological control is the potential and ecofriendly option. Biological pest control or utilization of natural enemies, in pest management is most successful to date, with hundreds of outstanding successes all over the world, and is undoubtly also the most promising alternative for the forcible future. According to DeBatch (1964) biological control when works, can solve permanent problem of pests.

Biological control is "the action of parasitoids, predators and pathogens in maintaining other organism's density at a lower average than would occur in their absence" (De Batch, 1964). Coppel and Martin (1977) defined Biological control as "Biological pest suppression in its narrow classical sense, usually restricted to the introduction by man of parasitoids, predators, and/or pathogenic microorganisms to suppress population of plants or animal pests".

Biological control can also be defined as, "the use or encouragement by man of living organisms or their productions for the population reduction of pest insects".

Biological control is an old practice of pest management. The first parasitic insect, *Cotesia* (*Apanteles*) *glomeratus* (L.) was recorded on cabbage white butterfly, *Pieris rapae* Boisduval larvae in 1602. Publications on biological control in 18th century refer to descriptions of wasps and flies emerging from other insects. Linnaeus suggested for the first time that aphids on plants can be controlled by using Ichneumonid parasitoid, *Ichneumonid aphidum* and also by the use of carabid beetle, *Calosoma sycophanta* against lepidopteran caterpillars in orchards. Similarly, Fisher (1965) controlled the citrus leaf eating insect by exposing predacious ant, *Ecophylla smaragdina* on citrus crop. Aristotle (384-322 B.C.) in his *Historia animalium* described the ravages of the wax moth of honey-comb and suggested that it bring disease into the swarm (Steinhous, 1956). In 19th century, Darwin suggested various parasitic insects to control number of economic pests (Coppel and Martin, 1977). About 110 pest species have been controlled by biological means in 16 countries, involving more than 225 projects (De Batch, 1964). Similarly, Simmonds (1970) reported 11 species of pests that have been controlled by using parasitoids and predators in some countries with collaboration of Commonwealth Institute of Biological Control. In 20th century several historical treatments, localization and broad range of biological control have published. Jonson (1957), Huffakar and Stinner (1971), Greathead (1957), Rao (1961), Rao *et al.* (1971), Sailer (1952), Hazen and Franz (1973), Coppel and Martins (1974, 1977), Nagarkatti (1981),Yeargan (1985), Cossentine and Lewis (1986), Lingren *et al.* (1988), Narasimhan and Chacko (1988), Sathe and Bhoje (2000), Sathe and Jadhav (2001), Sathe and Margaj (2001), Sathe *et al.* (2003), Sathe (2004), etc., worked on biological control of insect pests by using entomophagi.

The World's organizations like, CILB, CIBC and CABI are completely devoting to research in biological control as eco-friendly source of pest control. These organizations provide effective services to world agriculture to coordinate and administer biological control.

According to Castilachacon (1972) mass rearing and use of the entomophagous insects is a commercial practice against lepidopterous pests in South America, North America (Carpentier *et al.,* 1969, Ridgway, 1972), U.S.S.R (Dysart, 1973), China (Pu and Liu, 1962), India (Solayappan and Marar, 1974, Vardharajan, 1976, Sithnantham, 1977, Sathe and Bhoje, 2000, Sathe, 2004), Mexico (Humbert, 1976) and several other countries. At Bangalore, India, Commonwealth Institute of Biological Control (CIBC) was launched in 1957 to give fillip to the study of entomophagous insects. Later, several biocontrol centres have been established in India. Indian centres of Biological Pest Control such as Gorakhpur (U.P), Solan (H.P.), Hyderabad (A.P.) Shriganganagar (Rajasthan), Indian Agricultural Research Institute, Pusa (New Delhi), International Crop Centre Institute for Semi Arid Tropics, Patancheru (A.P.), Indian Agricultural Universities, VSI, Manjari, Pune and many traditional universities such as Shivaji University Kolhapur, Dr. Babasaheb Ambedkar Marathwada University Aurangabad, etc. are actively engaged in doing research on insect pest control by biological means.

In India, extensive survey for natural enemies of rice, sugarcane and coconut (Aphids, Rhinoceros beetles, etc.) have been made around 1955-60. Rao *et al.* (1971) reviewed biocontrol programmes of South East Asia. Many exotic biocontrol agents have been introduced against some of the more important agricultural pests.

Isotima javensis Rohw., a pupal parasitoid of the sugarcane topborer, *Tryporhyza nivella* (Fabricius) has been successively used by CIBC, Bangalore in biological control programme. Main Biocontrol Research Laboratory, Chengalpattu (Tamilnadu), used *Trichogramma* successfully against sugarcane internode borer. *Trichogramma* are being released at the rate of 1,20,000 to 2,00,000 adult parasitoids per acre for controlling the major Lepidopteran sugarcane borers. The Ichneumonid, *Campoletis chlorideae* Uchida is an important parasitoid of young larvae of *Helicoverpa armigera* (Hubn.) and 80 per cent parasitism have been recorded at Pipariya (Madhya Pradesh) (Shankaran and Nagarkatti, 1982) and 60 per cent on *Spodoptera litura* Fab. at Kolhapur (M.S.) (Sathe, 1987). The above species is well established in most of the states of India.

The term "parasitoid" was first used by Reuter (1913) to describe a life history, intermediate between that of predator and the true parasite. The parasitoids are different from true parasites by having host scattered only in class insecta and they always kill their hosts at the completion of their life cycle or end of the association. In other words, parasitoids are entomophagus.

The parasitoids are scattered chiefly in the Orders, Hymenoptera, Strepsiptera and Diptera. The Order Hymenoptera ranks first. Thousands of parasitic wasps frequently kill the pest populations. Ichneumonidae, Braconidae, Chalcidae, Trichogrammatidae, etc. are important parasitic families of the Order Hymenoptera. However, Ichneumonidae is widely used in biological control of insect pests (Sathe, 2004) which contains 60,000 described species in the world and rank first in hymenoptera (Gupta, 1988). Ichneumonids are exclusively parasitic, reported on all the major Orders of the insects like Lepidoptera, Coleoptera, Diptera and rarely Hymenoptera.

Ichneumonids are with more petiolate abdomen, second recurrent vein and more shinning and larger than other parasitoids. A parasitoid should have good searching ability, high degree of host specificity preferences, high reproductive capacity in relation to that of host and good adaptability to a wide range of environmental conditions. Economical exploitation of the parasitoids will be possible and fruitful, if we have information on above aspects. Since last three decades, Indian Ichneumonids has received very little attention in India except the work of Towns *et al.* (1961), Sathe *et al.,* 2003, etc.

Biocontrol programmes will not progress without understanding taxonomy, biology and ethology. Biological studies will further be helpful for mass rearing and augmenting the Ichneumonids in biological control of insect pests. Keeping in view all above facts, the present project was undertaken from Western Maharashtra including Ghats (Figure 1).

Ichneumonidae is one of largest family of Hymenoptera. Out of 60,000 species of the world about 1600 are from orient or Indo-Australian region (Townes, 1969). From India about 8000 species of Ichneumonids are believed to occur, because of the condusive environment of agro and forest ecosystems.

Ichneumonids are larval and pupal parasitoids of holometabolous insect pests. After their larval/pupal development in pest body they emerge as matured larva/adult from the pest body and thus kill the pest in the process. Adults are vegetarians. They feed on honeydew or sweet products secreted by some homopterous insects (Sathe and Margaj, 2001). They destroy a large number of pests of economic importance of Agriculture, horticulture, floriculture and forests.

Ichneumonids which can be used as biocontrol agents in integrated pest management need more attention by workers from various fields. Due to lack of information on them they are not used as biocontrol agents on large scale. The present project will add great relevance on precise taxonomic identity, biology, reproduction potential, parasitism aspects and biocontrol potential of some Ichneumonids from Western Maharashtra including Ghats.

As an Indian Ichneumonids Cameron at the end of 19th century made effective foundation which is followed by Morley (1912, 1913). Further more contribution has made by Cushman (1934), Beeson and Chatterjee 1935 and Rao (1953). Prof Gupta catalogued the Indian Ichneumonid fauna (Townes *et al.,* 1961). Prof Gupta also contributed on Indian Ichneumonids with his co-workers *viz.* Kamath and Gupta (1972), Jonathan and Gupta (1973), Gupta and Tikar (1976), Gupta and Gupta (1983), Gupta and Maheshwary (1977), Chandra and Gupta (1977) etc. This was followed by Kaur and Jonathan (1979), Nikam (1980), Jonathan (1980), Gupta (1987), Sathe *et al.* (2001) etc.

As regards to the biology, parasitism, intrinsic rate of increase, ethology of Ichneumonids Thompson (1941), Fisher (1959), Leius (1961), Tikar and

Thakre (1961) Gangrade (1964), Leong and Oatman (1968), Lingren *et al.* (1970), Oatman and Platner (1974), Chundurwar (1975), Schmidt (1975), Basarkar and Nikam (1981), Nikam and Gosawi (1982), Sathe (1990) etc. made significant contribution.

2

Materials and Methods

Pest management is essential part of modern agricultural practices for higher yields in the crops. More efficient pest control methods are continuously being searched and adopted, since a minute ultration in technique can have drastic results, both positive and negative (Patana, 1975). For biocontrol method continuous breeding stocks of both the pests and parasitoids are required for an uninterrupted study. Standard rearing techniques and best environment counts the success of biological pests control method. Hence, following materials and methods were adopted in the present study.

Materials

Glass Cages (Figures 2 and 3)

Two types of glass cages (Size, 25 × 25 × 30 cm) were used for rearing of host and parasitoids. Both cages were quadrangular in shape. Each type consists of wooden box and glass walls on three sides. In I type, the fourth side was closed by muslin cloth with a sleeve for handling the insects (Figure 2) and in the II type, fourth side of cage consist a glass window (Figure 3).

Figure 2: Glass Cage.   Figure 3: Glass Cage.

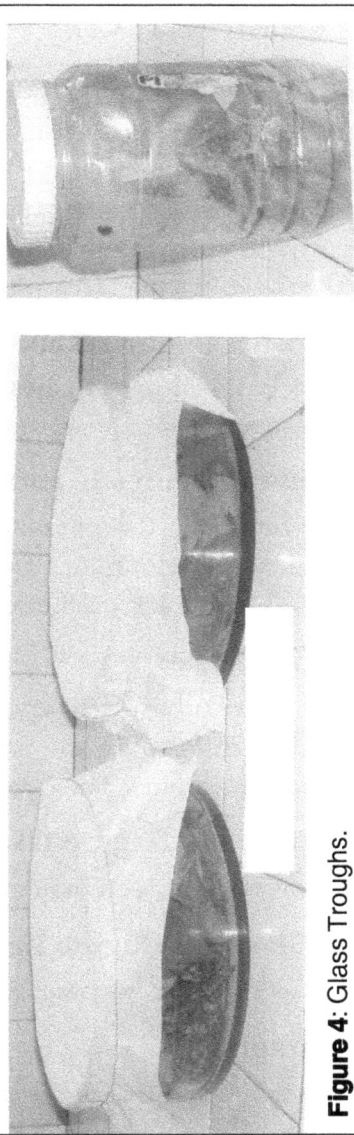

Figure 5: Plastic Container

Figure 4: Glass Troughs.

Glass Troughs (Figure 4)

Two sized glass troughs *viz.* 9 × 12 cm in height and 20 and 25 cm in diameter were used for rearing of caterpillars which were collected from the fields and also for screening pest parasitoids. After keeping larvae, the glass troughs were covered with muslin cloth.

Plastic Containers (Figures 5, 6 and 7)

Plastic containers of four different types (Size, 6.5 × 8, 5 × 6.2, 5 × 5 and 4 × 4 cm) were used for rearing the hosts and their parasitoids. The plastic lids of all containers were perforated for ventilation. For avoiding overcrowding or cannibalism the small size (4 × 4 cm) containers were used for keeping the larvae separately.

Petri Dishes (Figures 8 and 9)

Two types of petri dishes (Size, 18.5 cm and 9 cm in diameter) were used for rearing immature stages of pests and parasitoids.

Test Tubes (Figures 10 and 11)

For handling the parasitoids and pests and studying mating and oviposition behaviours in parasitoids three types of test tubes (Size, 19 × 2.5, 15 × 2, and 14.5 × 2.8 cm) were used.

Specimen Tubes

Specimen tubes of 10 cm and 5 cm in height and 2 cm in diameter were used for keeping the parasitoid cocoons for adult emergence and handling parasitoids. The open ends were closed with muslin cloth.

BOD Incubator (Figure 12)

It was used for eco-biological studies.

Insect Store Box (Figure 13)

It was used for keeping insects.

Photography

The whole mounts of the parasitoids and the various other parts were considered for photography.

Figure 8: Petri Dish

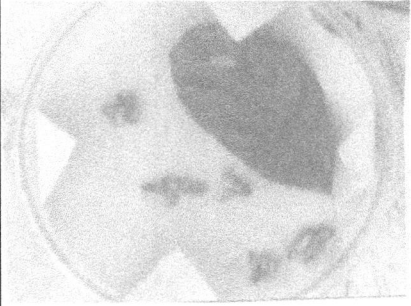

Figure 9: Petri Dishes

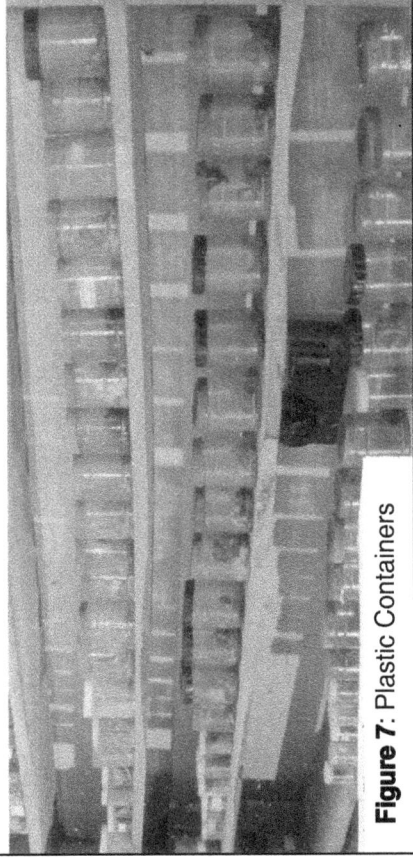

Figure 6: Plastic Containers

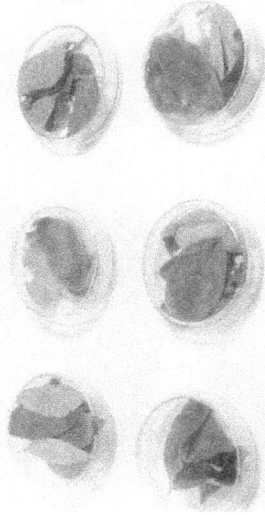

Figure 7: Plastic Containers

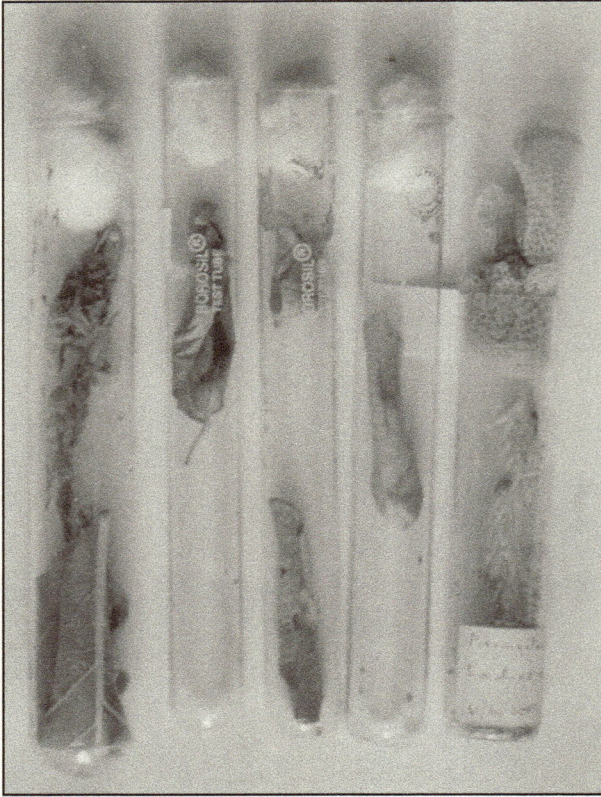

Figure 10: Test Tubes

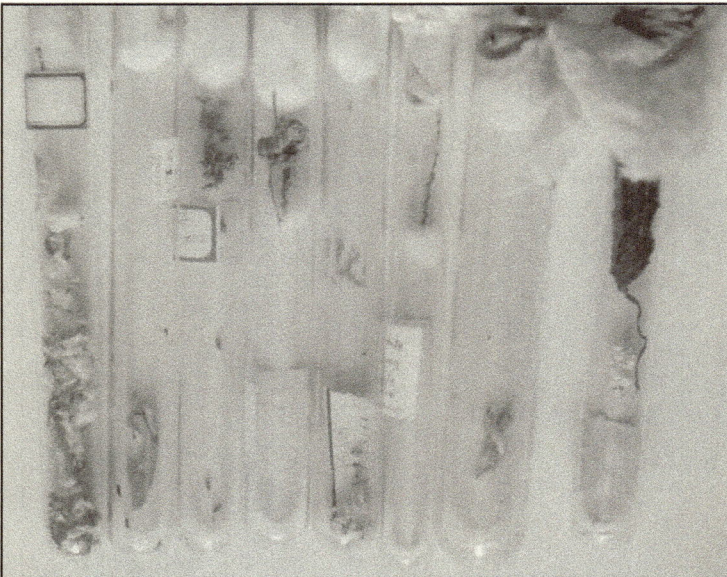

Figure 11: Test Tubes

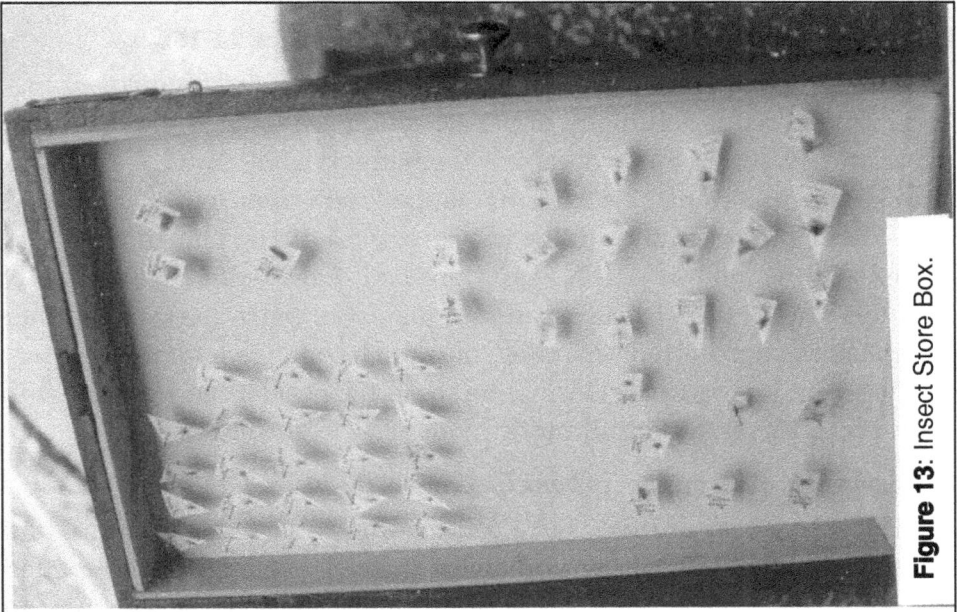

Figure 13: Insect Store Box.

Figure 12: BOD Incubator.

Methods

I. Rearing of Pest Species

1) Rearing of *Helicoverpa armigera* (Hubn) (Figures 14-16)

The larvae of *H. Armigera* were collected from the fields of gram/ jowar and reared individually in plastic containers (Size, 4 × 4 cm) up to pupation and adult emergence. Newly emerged pairs (♂ and ♀) of moths were kept in glass jars (29 × 9.5 cm) wrapped with tissue paper internally. With the help of muslin cloth the jars were closed. The muslin cloth was soaked with water, on which eggs were laid. Eggs were collected daily with the help of camel hair brush and placed in petri dishes with filter paper. Larvae transferred separately into small containers for their further development up to pupae. During the study the pest larvae were fed with gram pods/leaves and adult moths with 50 per cent honey.

2) Rearing of *Leucinodus orbonalis* Guene (Figures 17-19)

The larvae of *L. orbonalis* were collected from the fields of brinjal and reared individually in plastic containers (Size, 4 × 4 cm) up to pupation and adult emergence by providing Brinjal fruits as food. Newly emerged moths (pairs) (♂ : ♀) were kept in glass jars (29 × 9.5 cm) wrapped with tissue paper internally. The muslin cloth soaked with water was used for closing the jars on which eggs were laid by the moths of *L. orbonalis*. Eggs laid on muslin cloth were collected daily by camel hair brush and placed in petri dishes on filter paper for further development. Larvae transferred separately into small containers for their further development up to pupae. During the study the host larvae were fed with Brinjal fruits and adult moths with 50 per cent honey.

3) Rearing of *Thiocidas postica* Walker (Figures 20-23)

This species was reared under the laboratory conditions (25 + 1°C, 70 – 75 per cent R.H., 12 hr photoperiod). The larvae of *T. postica* were collected from the Ber fields of Kolhapur and reared individually in plastic containers for pupation and adult emergence. For mating purpose newly emerged pair of moths (1 male and 1 female) was caged in glass jar. After mating, females were separated from jars and placed in oviposition jar. The female started laying eggs on internally wrapped tissue paper in jar and on the wet muslin cloth of jar which was soaked in water. Eggs were collected by camel hair brush and placed in Petri dishes on filter paper. In each

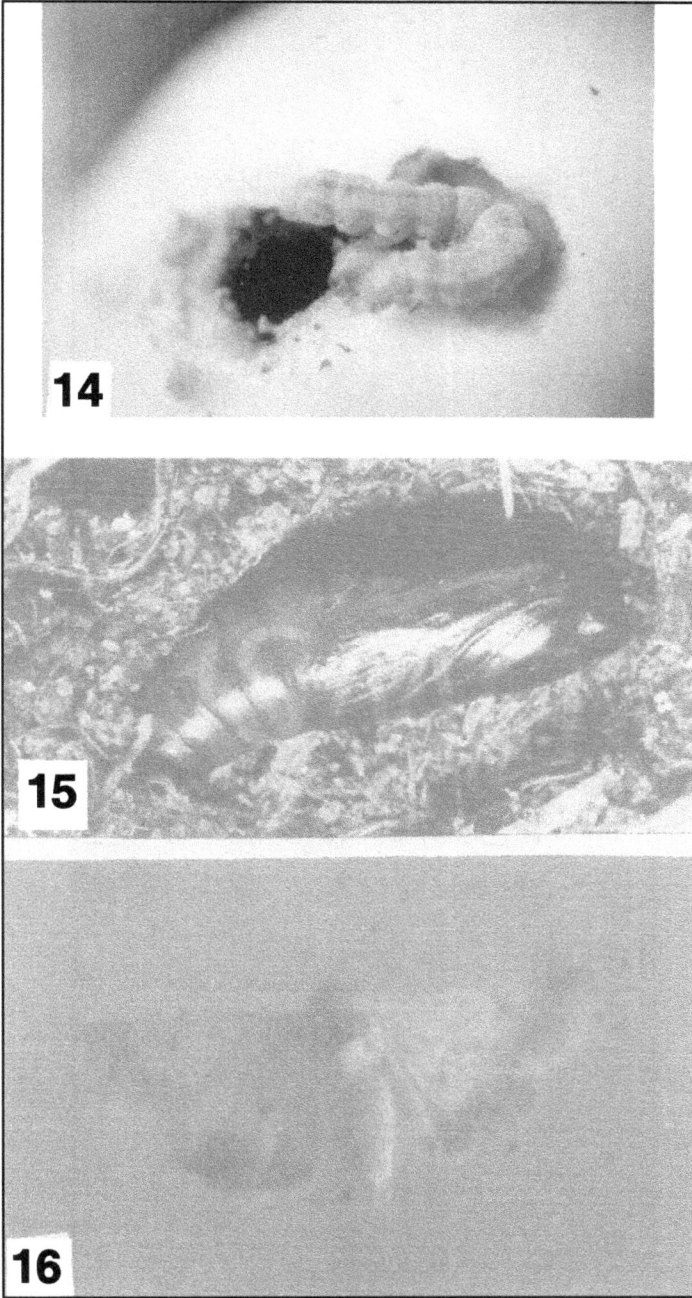

Figure 14–16: *Helicoverpa armigera.*
14: Larva, **15:** Pupa, **16:** Adult.

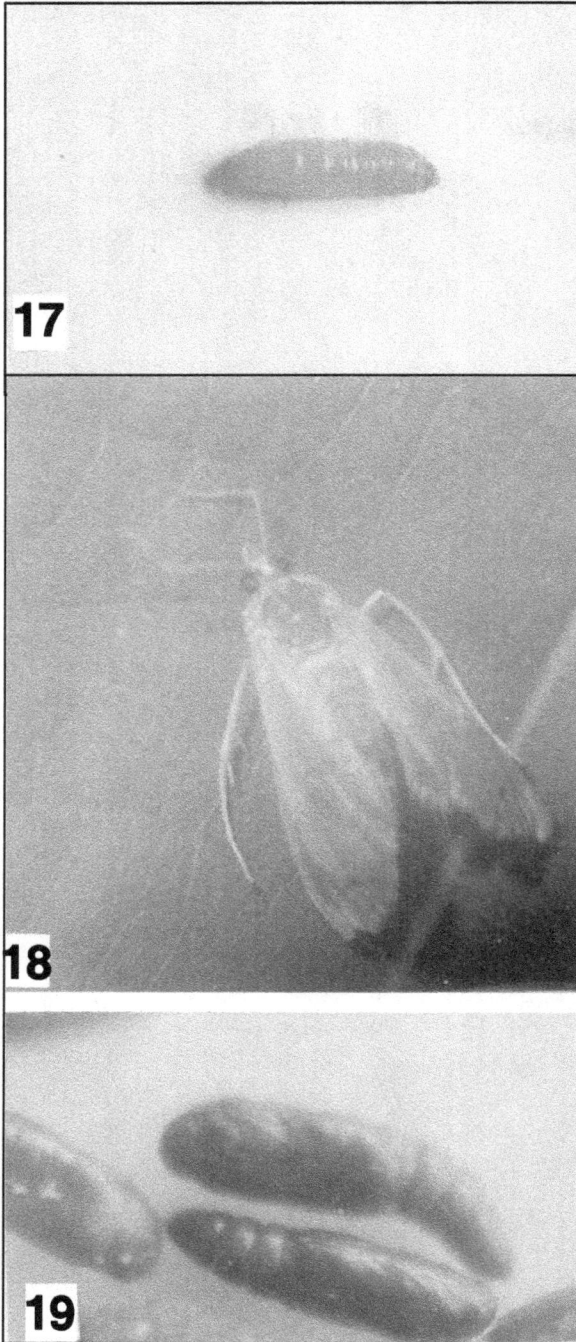

Figure 17–19: *Leucinodus orbonalis.*
17: Larva, **18:** Pupae, **19:** Adult.

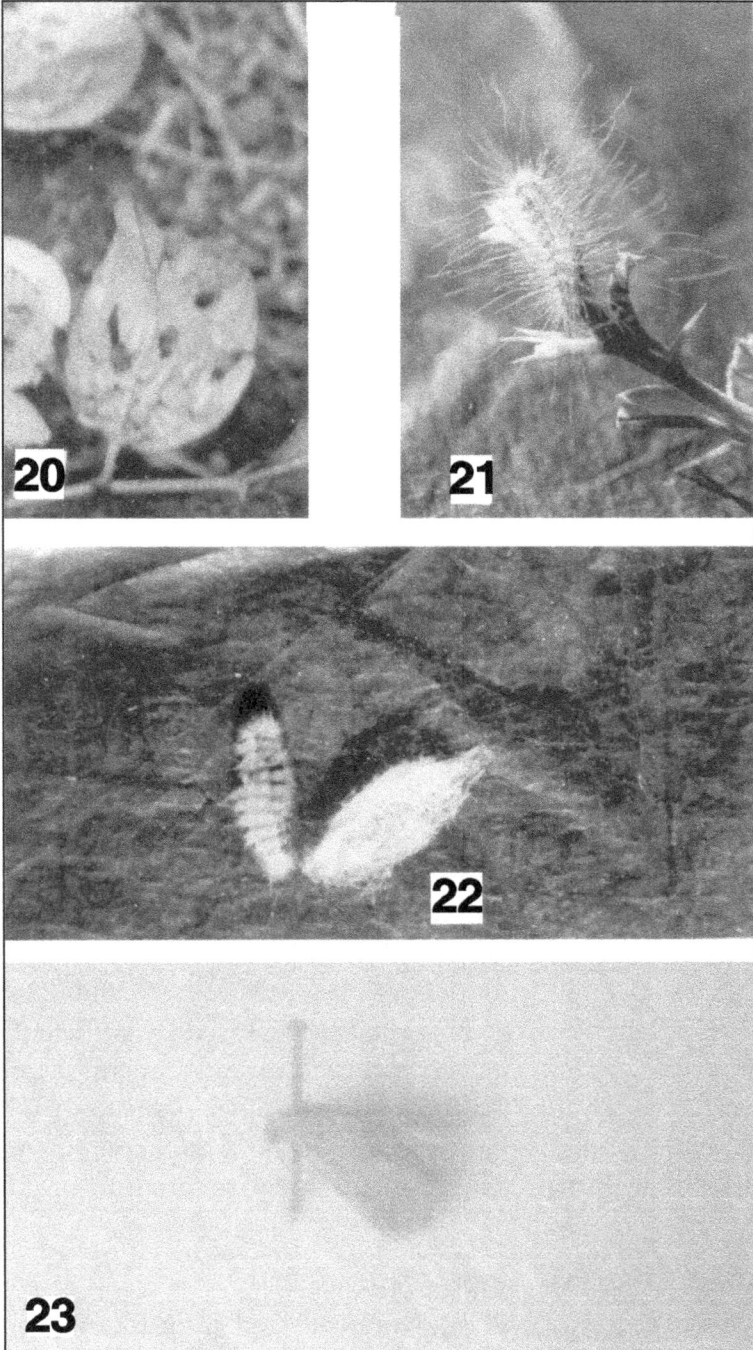

Figure 20–23: *Thiocidas postica.*
20: Larvae, **21:** Larva, **22:** Pupa; **23:** Adult.

container, 50 newly hatched larvae were placed by fine hair brush. Later, instars were transferred separately into small containers. The larva attained its full matured stage within 17 days and spins loose cocoon for pupation. The full grown larva moults into pupa within 3-4 hours. Pupa was obtect type and dark brown in colour. The pest larvae were fed with leaves of Ber (*Zizyphus jujuba* Mill) and adult moths with 50 per cent honey.

II. Rearing of Parasitoid Species

1) Rearing of *Pristomerus testaceous* Morley

The larvae of *L. orbonalis* were collected from Brinjal fields of Kolhapur to maintain the laboratory culture. Thirty early second instar larvae of *L. orbonalis* were exposed to five mated females of *P. testaceous* in an oviposited unit. Parasitoid eggs and larvae were collected after 12 hr interval, dissecting parasitized host larvae in normal saline solution for life cycle studies. The larval stages were treated with 50 per cent chloroform and 50 per cent ethanol and mounted in Hoyer's medium on micro slides after staining with acetocarmine. Instars were identified by observing size of head, capsule and mandibles (Short, 1970).

Laboratory culture of *L. orbonalis* was used for rearing *P. testaceous*. 30 hosts were exposed to five mated females of *P. testaceous* for oviposition in glass cage (25 × 25 × 25 cm) for 1 hr and later, parasitized larvae were removed from the cage and reared separately in plastic containers for development of parasitoids. Within 16-18 days parasitoid larvae fully developed and emerged from host body by breaking the body wall and killing the host species. The last instar of parasitoid pupate in plastic container/glass petri dish. Pupal period was 6 days. Within 22 days life cycle was completed. By repeating the above procedure sufficient number of parasitoids have been obtained and used for experiments. More or less same procedure was adopted for rearing of other parasitoids with some ultration mentioned under each species. Cocoons of parasitoids were also collected frequently from the field for supplementation of culture of parasitoids.

2) Rearing of *Diadegma fenestralis*

The cocoons of *D. fenestralis* were collected from the fields of jowar and reared in the laboratory for adult emergence. Cocoons were reared separately for avoiding mixing of sexes. Newly emerged parasitoids (1♂ and 1♀) were caged into test tubes (Size, 15 × 2 cm) for their mating. The mating was followed immediately after caging of sexes in test tubes. Five female parasitoids were exposed to 40 second instar larvae of the host in insect cages. With the help of the fine hair brush, the host larvae were introduced in cage through sleeve. One or other parasitoids quickly oviposited into host larvae offered and thus more parasitized host larvae were obtained within short time. In addition, parasitization was also made in test tubes (Size, 18 × 2.5 cm). After parasitism, host larvae were removed and kept separately in containers/petri dishes for further development and avoiding cannabalism. The parasitoids were fed with 50 per cent honey solution. This parasitoid was reared on *H. armigera* caterpillars.

3) Rearing of *Charops obtusus*

For the initial stock, the parasitoids were collected from the field. The individual cocoons were placed in each specimen tubes (Size, 10 × 2 cm and 5 × 2 cm). After adult emergence they were caged in a test tube for mating purpose. The mated females were used for oviposition on the pest larvae *i.e. T. postica*. Second instars of *T. postica* were exposed to mated female parasitoids for one hour. Only a single female was allowed to oviposit in one host to avoid super parasitism. Parasitized larvae were then reared in plastic containers by providing natural food material *i.e.* leaves of Ber. Within 21 - 22 days the parasitoid larvae emerge from parasitized *T. postica* larvae by breaking the body wall and immediately started spinning cocoons. Pupal period was 5-6 days. The parasitoid made circular cut at the posterior side of the cocoon and emerged outside. Thus, sufficient number of parasitoids were reared and further used in different experiments of the present study. The live cocoons were temporarily stored at 13°C for their longer use in experiments.

Identification of Parasitoids

Extensive collection/survey of Ichneumonid flies have been made from the districts, Kolhapur, Pune and Satara of Maharashtra state (Figure 1) from the years 2009 – 2013 from various agro ecosystems. The insects were collected early in the morning and evening. Many times, parasitized larvae of Lepidoptera and cocoons of Ichneumonids were also collected from the fields on host plants and reared in the laboratory. Few specimens were preserved by killing with chloroform and kept in 70 per cent alcohol/ 4 per cent formalin. Ichneumonids were pinned through mesothorax and kept in insect box. The detailed record on the species on locality, date of collection, name of collector and identification were made. The head, antenna, propodeum, wings, legs, tergites, ovipositor and ovipositor sheath were taken into account for identification of species. Measurements of the specimen were taken with the help of ocular micrometer in mm and calculated with the help of gaduated mechanical stage. Ichneumonid parasitoids were identified by consulting appropriate literature, Morley (1913), Townes *et al.* (1961) and Constantineanu *et al.* (1977), etc.

Distributional Record

Distributional record was made by observing and collecting the Ichneumonids from various places of three districts, Kolhapur, Pune and Sangli of Maharashtra (Figure 1) at 15 days interval by one man one hour search method.

Host Record

Host record of parasitoids was made by collecting parasitized pest species (hosts) and rearing them under laboratory conditions (25±10°C, 70–75 per cent R.H., 12 hr photoperiod) for the adult emergence and identification. A large number of references were consulted for confirming results of present work.

3

Collection, Identification and Abundance of Ichneumonids

Introduction

Ichneumonids (Order - Hymenoptera) are very potential biocontrol agents of lepidopterous and coleopterous pests. They parasitize larval and pupal stages of pests mostly and cause mortalities in them. Therefore, Ichneumonids are widely used in biological control of insect pests in India and other countries (Debatch, 1964; Sathe and Bhoje, 2000). 60 per cent successful programmes of biological insect pest control are from parasitoid categories. The rearing of parasitoids is also comparatively easy than predators. Therefore, hoping maximum utility of Ichneumonids in biological pest control the present work was carried out from Kolhapur district. The review of literature indicates that biodiversity of Ichneumonids has been attempted by Morley (1913), Townes *et al.* (1961), Sathe *et al.* (1996 a,b), Sathe (1986 a,b; 1987 a,b; 1992), Sathe *et al.* (2003) etc.

Materials and Methods

Biodiversity of Ichenumonids has been studied by collecting insect pest stages from three centres of collection spots namely Chandgad,

Gadhinglaj and Hatkanangale of high, mid and low rainfall respectively at 15 days interval from the years 2008-2010. The pest stages collected have been reared in the laboratory (24±1°C, 75 per cent R.H., 12 hr photoperiod) on their natural food for screening.

One man one hour search method was adopted for collection of Ichneumonids. The Ichneumonids have been identified by consulting Townes *et al.* (1961), Sathe *et al.* (2003).

Results

Results are recorded in Table 1. The results indicate that *C. chlorideae* and *Xanthopimpla* spp. were more common in all three regions of the district Kolhapur and were related with the certain crops (Table 1).

40 species of Ichneumonids have been collected and identified on various insect pests and results are presented in Figures 24-39 and research papers and National Conferences/Workshops. *Diadegma trichoptilus* (Cameron) was recorded on *Exelastis atomosa* (Wals.) (Lepidoptera), *Diadegma argenteopilosa* (Cameron) on *Spodoptera litura* (Fab.) (Lepidoptera), *Campoletis chlorideae* (Uchida) on *Helicoverpa armigera* (Hubn.) (Lepidoptera), *Eriborus argenteopilosa* (Cameron) on *Helicoverpa armigera* (Hubn.) (Lepidoptera), *Enicospilus* sp. on *Spodoptera litura* (Fab.) (Lepidoptera), *Xanthopimpla stemmator* (Thunburg) on *Chilo partellus* (Swin.) (Lepidoptera), *Xanthopimpla pedator* on *Bombyx mori* (L.) (Lepidoptera), *Diadegma surendrai* Gupta on *Phthorimaea operculella* (Zeller) (Lepidoptera), *Ecthromorpha notulatoria* Fab on *Mythimna separata, Charops* spp., on *Thiocidas postica* and *E. scintilans, Campoplex* spp. on *P. operculella, Henicospilus* sp. on *Euproctis fraturna, Diadegma fenesrtralis* on *H. armigera, Diadegma insulari* on *Plutella xylostella, Netelia* on *H. armigera, Pristomerus testaceous* on *Leucinodes orbonalis* Guene, *Charops obtusus* Morley on *Thiocidas postica* Walk and *Diadegma fenestralis* on *Helicoverpa armigera* (Hubn), *Charops charukeshi* on *Virachola* etc. The above some more important Ichneumonids found in Kolhapur, Satara and Pune region including Ghats. Results are communicated/published in journals.

Discussion

Sathe (1986a) made survey of Hymenopterous parasitoids on *Chapra mathias* Fab., a paddy pest in Kolhapur. He reported three hymenopterous

Table 1: Diversity of Ichneumonid Parasitoids from Kolhapur District.

Sr.No.	Parasitoids Species	Pest Species	Crop	Distribution	Occurrence
1.	Banchopsis ruficronis Cameron	Helicoverpa armigera (Hubn.)	Gram	Chandgad, Gadhinglaj, Hatkanangale	Rare
2.	Charops charukeshi S. and D.	Spilosoma obliqua (Walk.)	Pulses	Chandgad, Gadhinglaj, Hatkanangale	Common
3.	Charops patmangiri S. and D.	Thiocidas postica (Walk.)	Ber	Chandgad, Gadhinglaj, Hatkanangale	Rare
4.	Charops sp.	T. postica	Ber	Chandgad, Gadhinglaj, Hatkanangale	Common
5.	Charops sp.	Vilrachola iscolartes (Fabr.)		Gadhinglaj, Hatkanangale	Common
6.	Campoletis chlorideae Uchida	H. armigera Spodoptera litura (Fab.)	Groundnut	Gadhinglaj, Hatkanangale	Common
7.	C. chlorideae	H. armigera Spodoptera exigua (Hubn.)	Groundnut	Chandgad, Gadhinglaj, Hatkanangale	Common
8.	Diadegma fenestralis (Cameron)	H. armigera	Jowar	Gadhinglaj, Hatkanangale	Common
9.	D. trichoptilus (Cameron)	Exelastis atomosa (Wal.)	Red Gram	Gadhinglaj, Hatkanangale	Common
10.	D. trochanterata Morlay	Dichocrocis puntiferalis (Guence)	Mulberry	Gadhinglaj, Hatkanangale	Common
11.	D. vulgari Morlay	S. exigua	Jowar	Gadhinglaj, Hatkanangale	Common
12.	D. recini R. and K.	D. puntiferalis	Caster	Gadhinglaj, Hatkanangale	Common
13.	Diatora lissonota Viereck	Achea janata Linn.	Caster	Chandgad, Gadhinglaj, Hatkanangale	Rare
14.	Ecthromorpha sp.	Mythmna separata (Walker)	Jowar	Gadhinglaj, Hatkanangale	Common
15.	Eriborus trochanteratus (Morley)	H. armigera, P. operculella	Potato	Hatkanangale	Common
16.	E. sinicus Holm.	Tryporiza insertulus Wlk.	Sugarcane	Gadhinglaj, Hatkanangale	Common
17.	Enicospillus sp.	H. armigera, S. litura	Jowar	Chandgad, Gadhinglaj, Hatkanangale	Common
18.	Gonyphus chaitshri S. and D.	Erias vitella Fab.	Cotton	Gadhinglaj, Hatkanangale	Rare

Contd...

Table 1–*Contd...*

Sr.No.	Parasitoids Species	Pest Species	Crop	Distribution	Occurrence
19.	*G. nursei* (Cam.)	*E. vitella*	Cotton	Gadhinglaj, Hatkanangale	Common
20.	*Goryphus* sp.	*E. vitella*	Cotton	Hatkanangale	Rare
21.	*Itoplectes narangae* Ashmead	*Chilo suppressalis* (Hampson)	Paddy	Chandgad, Gadhinglaj, Hatkanangale	Common
22.	*Isotima javensis* Rohwer	*Acigona steniella*	Sugarcane	Hatkanangale	Rare
23.	*Netelia ephippata* Smith	*A. janata*	Caster	Gadhinglaj, Hatkanangale	Common
24.	*Netelia* sp.	*A. janata*	Caster	Gadhinglaj, Hatkanangale	Common
25.	*Netelia*	*M. separata*	Jowar	Gadhinglaj, Hatkanangale	Common
26.	*Perilissus cingulator* (Morley)	*Athalia proxima* (Klug)	Mustard	Hatkanangale	Common
27.	*Pristomerus testaceous* (Morley)	*Leucinodous orbonalis* Guenee	Brinjal	Chandgad, Gadhinglaj, Hatkanangale	Common
28.	*Pristomerus euzopherae* Viereck	*Euzophera perticella* Rag.	Sweet potato	Gadhinglaj, Hatkanangale	Common
29.	*Pristomerus valnerator* Pan.	*Phthorimaea operculella* Zeller	Sugarcane	Gadhinglaj, Hatkanangale	Common
30.	*Xanthopimpla cera* Cam.	*Scirpophaga nivella* Fab.	Sugarcane	Chandgad, Gadhinglaj, Hatkanangale	Common
31.	*X. nursei* Cameron	*Sylepta derogata* Fab.	Cotton	Gadhinglaj, Hatkanangale	Common
32.	*X. punctata* F. *X. pedator* (Linn.)	*Chilo partellus* (Swin.) *C. partellus*	Jowar	Gadhinglaj, Hatkanangale	Common
33.	*X. stemator* Cameron	*C. partellus*	Jowar	Gadhinglaj, Hatkanangale	Common
34.	*Xanthopimpla* sp.	Paddy borers	Paddy	Chandgad, Gadhinglaj, Hatkanangale	Common
35.	*X. transversalis* V.	Jamun borer	Jamun	Chandgad, Gadhinglaj, Hatkanangale	Common
36.	*Ichnojoppa* sp.	*Parnara mathias*	Paddy	Chandgad, Gadhinglaj, Hatkanangale	Common
37.	*Trathala flavor-orbitalis* Cameron	*C. partellus*	Jowar	Hatkanangale	Common
38.	*Cremastus hepaliae* Cush.	*P. operculella*	Potato	Hatkanangale	Rare
39.	*Diadegma surendrai* Gupta	*P. opercullella*	Potato	Hatkanangale	Common

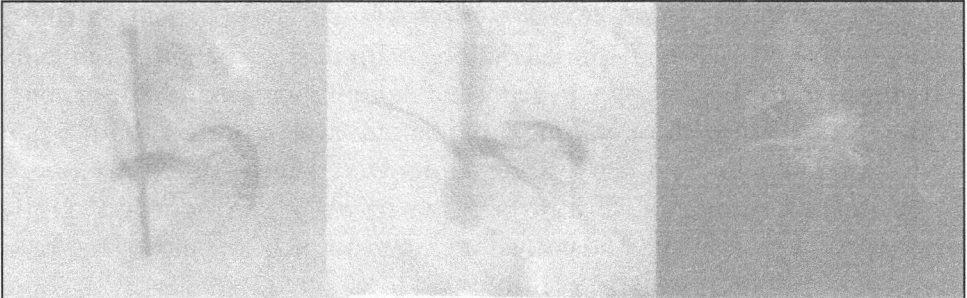

Figure 24: *Enicospilus* sp.　**Figure 25**: *Netelia* sp.　**Figure 26**: *Enicospilus* sp.

Figure 27: Unidentified.　**Figure 28**: *Charops obtusus.*　**Figure 29**: *C. charukeshi.*

Figure 30: Unidentified.　**Figure 31**: Unidentified.　**Figure 32**: *Charops obtusus* .

Figure 33: *Diadegma .*　**Figure 34**: *Diadegma*
　　　　trochanterata　　　　　　*fenestralis.*

parasitoids namely *Apanteles baoris, Ichnojoppa* sp. and Chalcid. Further, he reported that *A. baoris* appeared along with pest and about 5 per cent parasitism in the beginning was recorded, later it increased to 87 per cent. While, parasitism with remaining 3 parasitoids was reported about 5 per cent. From a single host larva 20-53 adults were successfully developed. In 50 hosts average of 43 adults of *A. baoris* were recorded. High percentage of parasitism good parasitoid range might be of help in checking the pest population in field conditions. However, *C. mathias* is a good example of multiparasitism and superparasitism in all 3 rainfall regions of Kolhapur.

Sathe (1986b) also made survey of natural enemies of *Exelastis atomosa* Wals. on pigeon pea in Kolhapur. He reported 3 parasitoids on *E. atomosa.* Out of which *D. trichoptilus* (Cameron) was an Ichneumonid parasizing about 15-18 per cent larvae. A survey of parasitoids of Ber hairy caterpillar *Thiocidas postica* Wlk. from Kolhapur have also been made by Sathe (1987a). He reported 3 parasitoids namely *A. creatonoti* Viereck, *Charops* sp. (Ichneumonidae) and *Tachina (Exorista) faliax* Meigen. The per cent parasitism in these parasitoids was 30 per cent, 15 per cent and 8 per cent respectively.

In a survey of natural enemies of *Spodoptera litura* (Fab.) from Kolhapur, Sathe (1987 b) reported 10 hymenopterous parasitoids parasitizing *S. litura*. The highest parasitization by *C. chlorideae* was 60 per cent in fields of groudnut. Early second instar caterpillars were preferred for parasitization. *C. chlorideae, Enicospilus* sp. and *Ecthromorpha* sp. were noted on *S. litura* as Ichneumonid parasitoids. On *Spodoptera exigua* 13 hymenopterous parasitoids have been recorded from Kolhapur. Maximum parasitization caused by *C. chlorideae* in 3 regions was 35 per cent, 60 per cent and 71 per cent in high, mid and low rainfall areas respectively. The lowest parasitization, 3 per cent was recorded by *Ecthromorpha* sp.

Recently, Sathe (1992) prepared index of hymenopterous parasitoids and pest insects from Maharashtra. Sathe *et al.* (2003) reported 13 species of Ichneumonids from Maharashtra. Similarly, Kavane *et al.* (2003) also reported two hymenopterous parasitoids namely *Xanthopimpla pedator* (Ichneumonidae) and *A. angaleti* (Braconidae) from Satara district during the years 2000 to 2003. Very recently, Sathe and Chougale (2006) reported ten hymenopterous parasitoids on *Helicoverpa armigera* (Hubn.). Out of which *C. chlorideae* was dominant over other parasitoids reported. The present work will be very good base line data for utilization of Ichneumonids in biological pest control programmes.

4

Life History of Parasitoids

Introduction

Ichneumonid wasps occur in thousands in the world, possessing complex and fascinating biologies (Doutt, 1959). They frequently determine pest population densities. Hence, widely used in biological pest control programme (De Batch, 1964; Sathe, 2004). Ichneumonids are entomophagous insects and are good friends of farmers as they kill the pest species in agro and forest ecosystems. Ichneumonids found attacking pests from orders Lepidoptera, Hymenoptera, Coleoptera, Hemiptera etc.

The review of literature indicates that Thorpe (1932), Short (1970), Juillet (1960), Gangrade (1964), Sathe (1988, 1990, 2008, 2012), Sathe and Dawale (1997), etc. studied life history aspects in Ichneumonids. The life cycle data is helpful in understanding the development, morphological peculiarities of the immature forms, authentic identity and in mass rearing of species. Therefore, the present topic is devoted for life cycle studies of *P. testaceous, C. obtusus, D. fenestralis,* attacking *L. orbonalis, T. postica,* and *H. armigera* respectively.

Materials and Methods

Cultures of parasitoids and their hosts were maintained in the laboratory (25 ± 1°C, 70-75 per cent R.H., 12 hr photoperiod) as mentioned in the chapter, materials and methods and used in life history studies of above parasitoids. Life cycle in all parasitoids was studied by exposing second instar (3-4 day old) host larvae to them. Host densities for parasitization by *P. testaceous* and *C. obtusus, D. fenestralis* were 30, 50, 40 and 30 host larvae respectively. The host densities mentioned above were exposed to respective parasitoids for 24 hr and sufficient number of parasitoid immature stages were collected for morphological observations and instar durations. The life cycle was studied by keeping the record of egg laying period, incubation period, larval period, pupal period and adult longevity. The experiments were replicated for 5 times.

Results

1. Life Cycle of *P. testaceous* (Figures 35A–D)

Egg (Figure)

The newly deposited eggs were white, thin walled and typically hymenopteriform. They were randomly deposited in the body cavity of the host larva. About 35-48 eggs were deposited per host. The eggs increased in the size after oviposition. The length and width of 20 eggs averaged 0.25 mm (range 0.20 to 0.30 mm) and 0.07 mm (range 0.05 to 0.09 mm) respectively. Its cephalic end has broadened. Incubation period was 3 days.

Instars

The parasitoid showed 5 instars. First two were caudate type and rests were hymenopteriform.

1st Instar (Figure 35B)

The 1st instar showed a translucent body, a broad head, 3 thoracic, 7 abdominal segments and a tapering tail. After 3 days body became more opaque and the head narrower. The body length of 20 individuals averaged 1.50 mm (range 1.35 to 1.75 mm) and width was 0.18 mm (range 0.14 - 0.22 mm). The head capsules of 20 individuals averaged 0.13 mm in length (range 0.10 - 0.15 mm) and 0.09 mm (range 0.08 - 0.10 mm) in width. In 20 individuals, mandibles averaged 0.024 mm (range 0.021 - 0.028 mm)

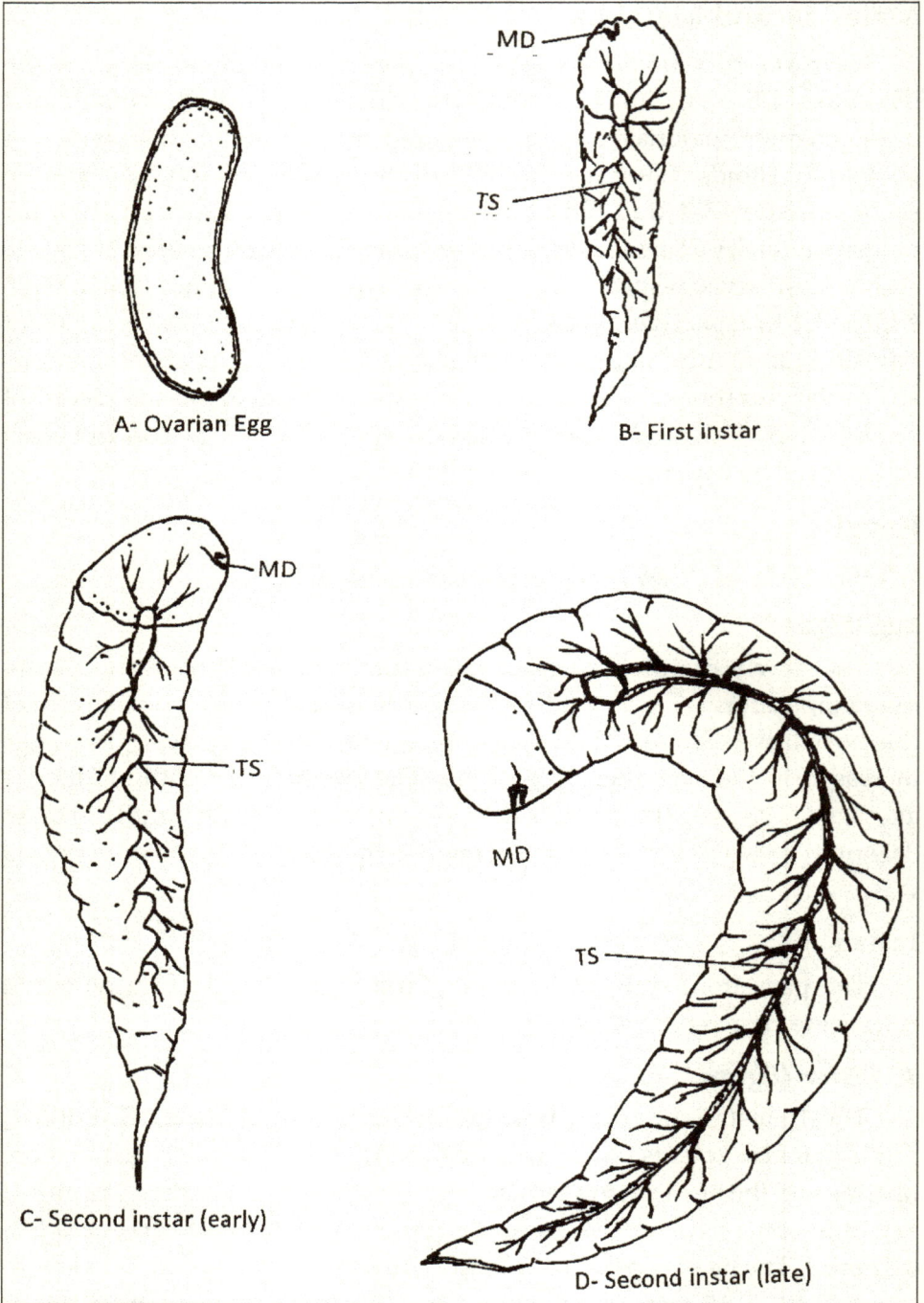

Figure 35: A–D: *P. testaceous.*
MD: Mandibles; TS: Tracheal system.

and 0.08 mm (range 0.007 - 0.012 mm) in length and width respectively. The caudate was transparent and tapering. The first instar lasted for 4 days.

IInd Instar (Figure 35C)

The body of second instar was cylindrical, slightly curved and white. The integument of larvae was very thin. The larva showed a narrow head, 13 well defined segments and a prominent tapering tail. No spiracle could be seen in second instars but tracheal system showed two lateral longitudinal trunks (Figure). The average length and width of the body measured in 20 individuals were 1.85 mm (range 1.75 - 1.95 mm) and 0.35 mm (range 0.32 - 0.38 mm) respectively. This stage lasted for 2 days.

IIIrd Instar

This opaque white instar lasted for 2 days and was without tail and was slightly curved. 20 individuals averaged 2.30 mm (2.10 - 2.40 mm) and 0.85 mm (range 0.80 -0.89 mm). Head capsule averaged, 10.0 mm (range 0.85 - 10.2 mm) in length.

IVth Instar (Figure 36E)

It was white opaque, well segmented, straight, considerably longer than previous 3 instars. In 20 individuals it was averaged 3.60 mm (range 3.50 to 3.75 mm) and Head capsule averaged 1.75 mm (range 1.073 to 1.79 mm). This stage lasted for 2 days. Mandibles averaged 0.55 mm in length and 0.40 mm in width. The larva consumed internal tissues of the host insect keeping vital organs intact.

Vth Instar (Figure 35F)

This instar was opaque yellow with no tail and slightly shorter and broader than fourth instar larva. 20 individuals averaged 3.40 mm and 0.75 mm in length and width respectively. Mandibles averaged 0.06 mm and 0.045 mm in length and width respectively. This stage lasted for 3 days. Fully matured larva came out by breaking the host body and spun cocoon.

Cocoon

Cocoon was dirty whitish. 4.00 mm and 1.10 mm in length and width respectively. Parasitoid emerged from cocoon by taking circular cut.

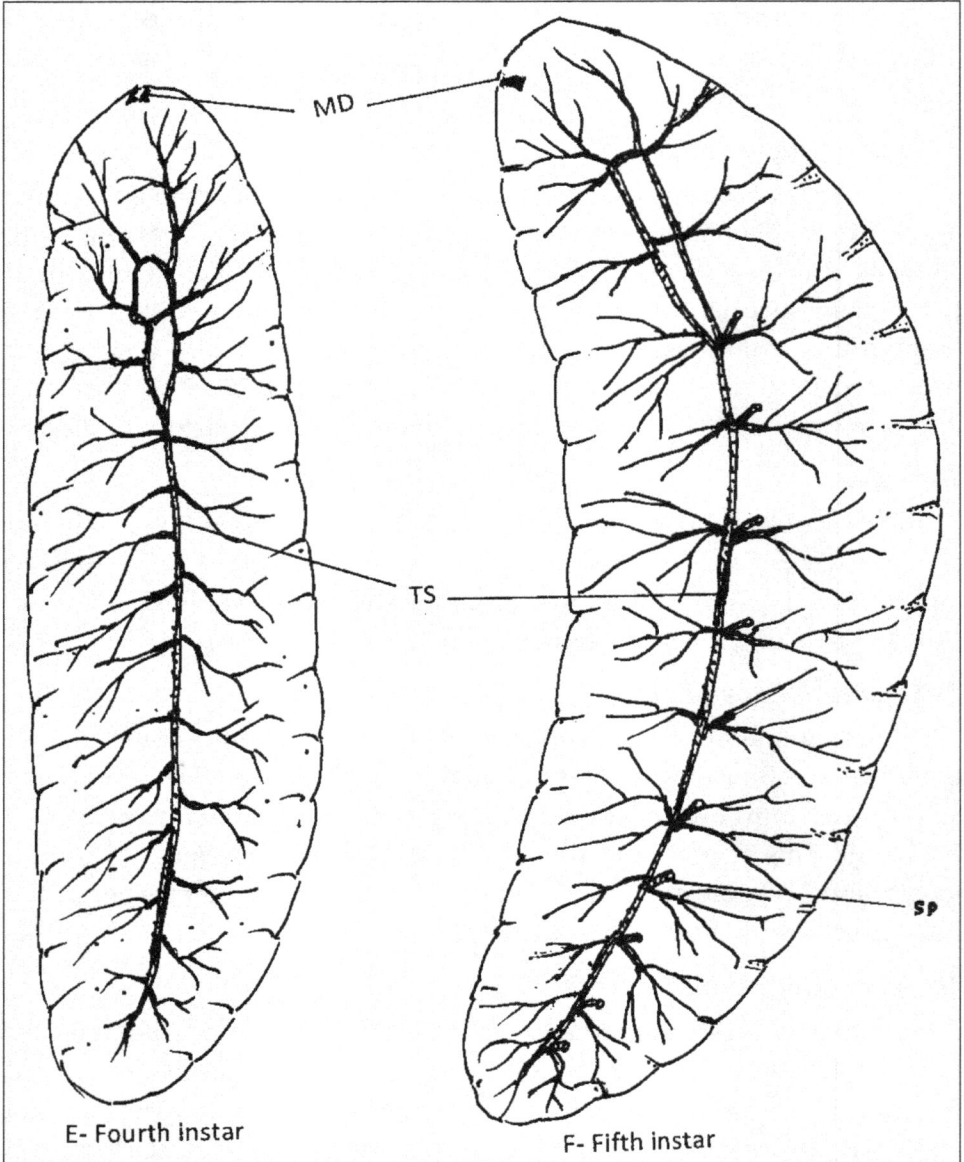

Figure 36: E–F: *P. testaceous.*

Pupa (Figure 37G)

Pupa was exarate type, brownish, 3.50 mm and 1.00 mm in length and width respectively. Pupal stage lasted for 6 days and life cycle completed within 22 days.

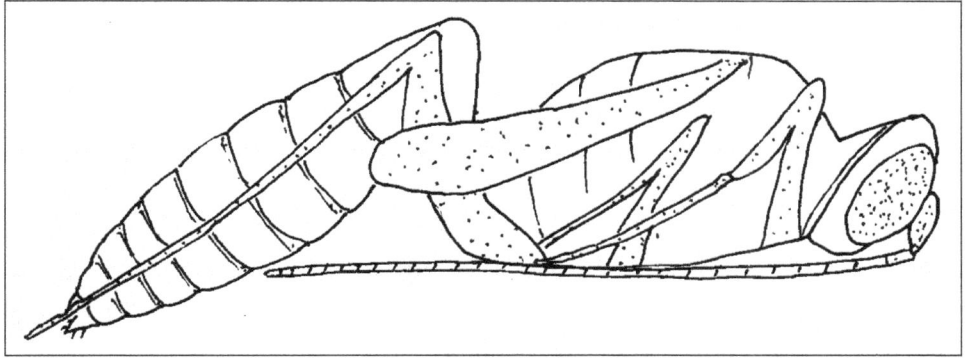

Figure 37: G: Pupa

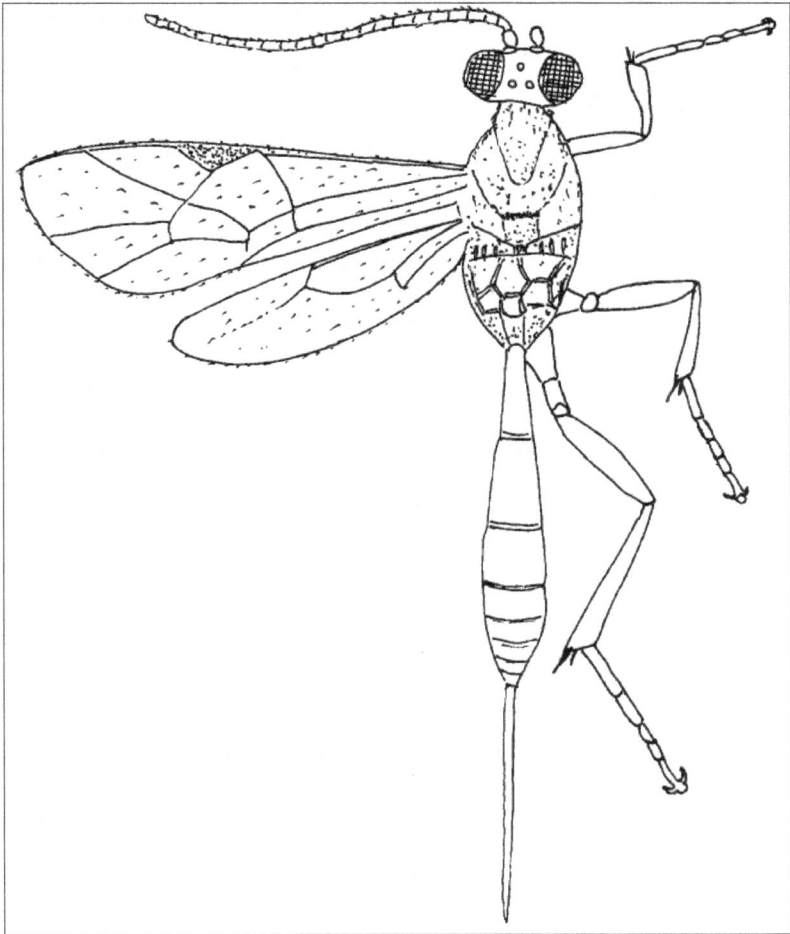

Figure 38: *P. testaceous*
H: Adult-Female.

2. Life Cycle of *C. obtusus*

Egg (Figure 39A)

The ovarian eggs were somewhat oval and elongated. The newly deposited eggs were white, thin walled and slightly curved. One or two eggs were deposited on the body of *T. postica* caterpillar but some times two or three have been deposited. The average length and width of 20 eggs were 0.36 mm (range 0.30 to 0.43 mm) and 0.010 mm (range 0.08 to 0.014 mm) respectively. The size of egg was increased after 48 hr. Eggs hatched within 3 days after oviposition.

Instars

In *C. obtusus* there were 5 larval instars. First three were caudate type, while and the last two instars were hymenopteriform.

I^st Instar (Figure 39B)

The first instar larva was creamy or opaque with 13 post cephalic segments and with characteristic long tapering tail. Segments were dissimilar. There was no differentiation between the thoracic and abdominal region but all the 13 segments were well defined. The tail was approximately a little longer than half the length of body, at the point from which the tail started. There exists a minute invagination of the integument. The tracheal system showed two longitudinal trunks on the body. No spiracles were observed. First instar larvae in 20 individuals averaged 2.00 mm in length (range 2.00-2.38 mm) and 0.28 mm in width (range 0.26-0.38 mm). The average length of 20 head capsules was 0.210 mm (range 0.215-0.225 mm) and width was 0.180 mm (range 0.178-0.190 mm). The length of 20 mandibles was averaged 0.026 mm (range 0.023 to 0.030 mm) and width was 0.020 mm (range 0.018 to 0.024 mm). This stage lasted for 4 days.

IInd Instar

IInd instar was opague white with apparently variable body shape and tail was considerably shorter than first instar. Complete disappearance of tail was noticed in the late form and was followed by thickening of body. The head and tail were considerably reduced while, the body found increased in size. The tail became equal to the size of the head. The second

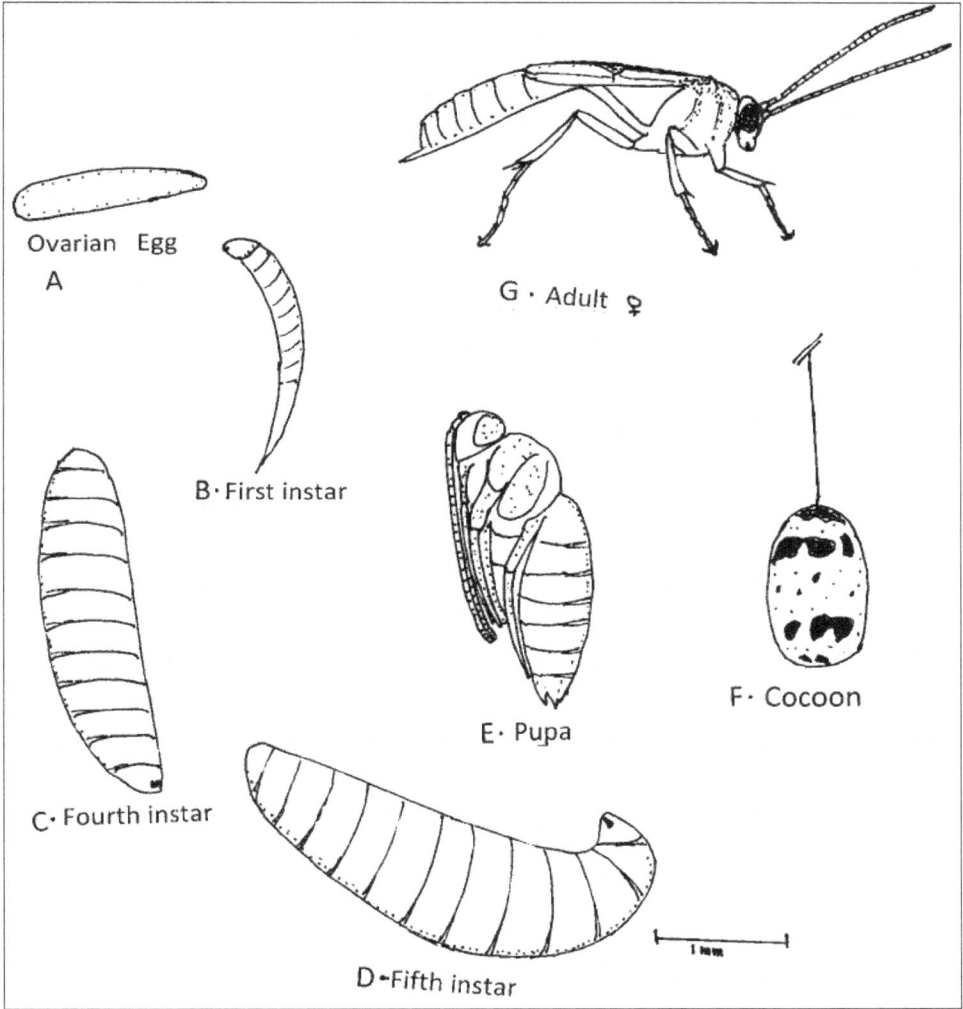

Figure 39: A–G: *C. obtusus.*

instar was different from first by its increased size, reduced tail, more sclerotized head and further branching of the tracheal system. The average body length of 20 individuals was 2.30 mm (range 2.25 to 2.40 mm) and width was 0.60 mm (range of 0.55 to 0.65 mm). The length and width of 20 head capsules averaged 0.36 mm (range 0.34 to 0.40 mm) and 0.243 mm (range 0.235 to 0.250 mm) respectively. In late instar tail was absent. The length of 20 mandibles averaged 0.039 mm (range 0.035-1.043 mm) and width 0.033 mm (range 0.030 to 0.037 mm). This stage lasted for 2 days.

IIIrd Instar

Third instar was also apaque white with clear and larger body segments and larger in size and shape than that of second instar. The tracheal system was well developed but spiracles were still not seen. Both the caudal and cephalic ends were similarly rounded but slightly tapered at the both ends. The head was more sclerotized. The average body length of 20 individuals was 4.25 mm (range 4.20 to 4.40 mm) and capsule in 20 individuals averaged 0.78 mm (range 0.75 to 0.80 mm) and 0.60 mm (range 0.58 to 0.65 mm) respectively. The length of 20 mandibles averaged 0.050 mm (range 0.048 to 0.058 mm) and width 0.036 mm (range 0.033-0.040 mm). This stage lasted for 2 days.

Vth Instar (Figure 39D)

Fifth instar was also hymenopteriform. Body segments were opaque white at first but became yellowish on reaching the maturity. The larva consumed all parts of host tissue except the alimentary canal. The average body length of 20 individuals was 5.75 mm (range 5.70 to 5.80 mm) and width was 1.32 mm (range 1.20 to 1.40 mm). In 20 head capsules the average length and width were 1.60 mm (range 1.55 to 1.70 mm) and 0.82 mm (range 0.80 to 0.90 mm) respectively. The length of 20 mandibles averaged 0.090 mm (range 0.086 to 0.098 mm) and width 0.065 mm (range 0.060-0.070 mm).

Tracheal system was well developed, with much more branches and with 10 pairs of delicate spiracles. The head structures were almost complete and strongly sclerotized. All the body segments were well marked. Full grown larva came out of the host body and spun cocoon. This stage lasted for 2 days.

Cocoon (Figure 39F)

Cocoons were barrel shaped, smooth and grayish coloured with irregular black spots. The length of 20 cocoon averaged 5 mm (range 4 to 7 mm) and width 3.5 mm (range 3 to 4 mm). Cocoons were with peduncles in field and found typically in hanging condition.

Pupa (Figure 39E)

The pupa was exarate type and at first it was creamy white. Later, it became yellow and at last dull black coloured. The emergence of adult occurred through a circular opening at the cephalic end. The pupal stage lasted for 4 days.

Adult (Figure 39G)

Female was 16.00 mm in body length. Head, thorax and last few segments of abdomen were black. Antenna, hind legs, femora and ovipositor were black.

3. Life Cycle of *D. fenestralis (Figures 40A–G)*

Egg

The eggs were elongately curved. The chorion was smoothly opaque and white. The mean length was 0.20 mm (range 0.18 - 0.25) and width was 0.050 mm (range 0.040-0.065). The female randomly deposited eggs in the host body. Incubahon period was 3 days (range 2.5 to 3.5).

Larvae

In *D. fenestralis* 5 larval instars were seen and were leg less.

Instars

First Instar (Figure 40B)

The first larval form is creamy or opaque white, and has 13 post cephalic segments and characteristic long, tapering tail. The body length of 20 individuals of first instar averaged 1.20 (range 0.90-1.5 mm) and width 0.158 mm (0.15-0.17 mm) head capsule averaged 0.150 mm in length (range 0.12 to 0.16 mm), and width 0.110 mm (range 0.1 to 0.130 mm). The length of mandibles in 20 individuals was 0.018 mm (range 0.017-0.02 mm) and mean width was 0.012 mm (range 0.011-0.013 mm). A considerable increase in size of larvae have been noticed within few hours after its hatching from the egg. This stage lasted for 3 days.

Second Instar (Figure 40C)

The second instar was opaque and white with the tail considerably shorter. Complete disappearance of tail was noticed in late second instar. The body segments were clear. The length of 20 individuals averaged 1.85 mm (range 1.47-1.90 mm) and width 0.250 mm (range 0.22 to 0.28 mm). In 20 head capsules the mean length and width were 0.210 mm (range 0.19-0.22 mm) and 0.180 mm (range 0.160-0.195 mm) respectively. The length of mandibles averaged 0.026 mm (range 0.025-0.028 mm) and width 0.018 mm (0.017-0.02 mm). This stage lasted for one day.

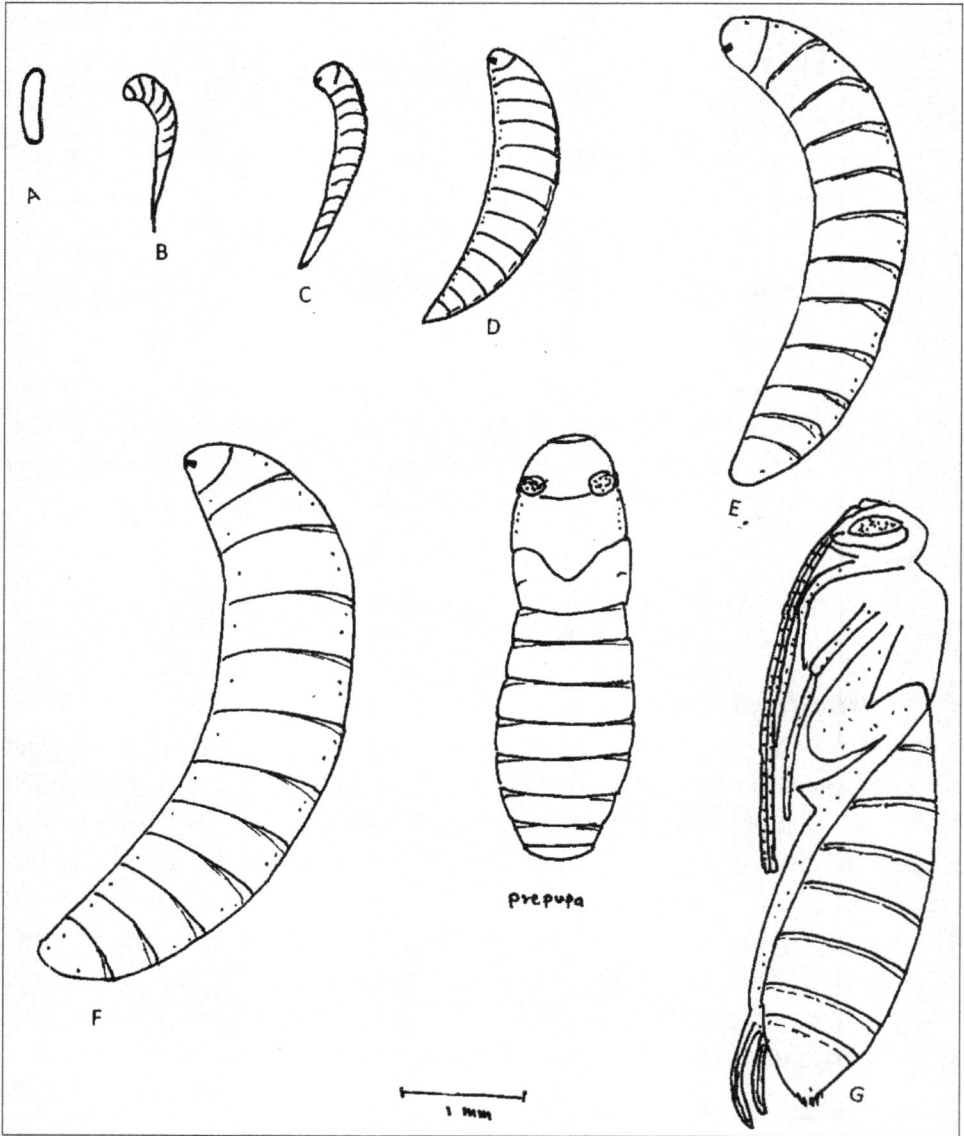

Figure 39: A–G: *D. fenestralis.*
A: Ovarian egg; B: First instar; C: Second instar (early); D: Third instar (late);
E: Fourth instar; E: Fifth instar; G: Pupa.

Third Instar

Third instar was opaque white and the body segments were clearly
seen. This instar was slightly larger in size and shape than that of second

instar. No tail was seen in this form. The average length and width of the body measured in 20 individuals were 3 mm (range 2.15-3.22 mm) and 0.50 mm (range 0.48-0.53 mm) respectively. Head capsule length and width averaged 0.40 mm (range 0.38-0.42 mm) and 0.33 mm (range 0.32-0.34 mm) respectively. The length of mandibles averaged 0.035 mm (ranged 0.034-0.036 mm) and width averaged 0.024 mm (range 0.023-0.026 mm). This instar lasted for only one day. The tracheal system was well developed but the spiracles were not clear.

Fourth Instar (Figure 40E)

This opaque white instar was similar to that of the previous stage. It was smooth, elongate and hymenopteriform. The larva was very active in feeding. It consumed the haemolymph, adipose tissue and malphigian tubules of the host. The mean length of body was 3.80 mm (range 3.5-4.10 mm) and width was 0.75 mm (range 0.74-0.76 mm). The head capsules averaged 0.70 mm (range 0.68-0.72 mm) and 0.40 mm (range 0.39to 0.42 mm) respectively. The mandibles averaged 0.065 mm (range 0.063-0.068 mm) and 0.032 mm (range 0.030-0.036 mm) respectively.

Tracheal system was well developed, nine pairs of spiracles were seen. The first pair was situated between 2nd and 3rd segments and one pair each from the 4th to 12th segments. This stage stayed for one day.

Fifth Instar (Figure 40F)

This instar was also hymenopteriform and opaque white at first, but became pinkish on reaching the maturity. In 20 individuals body length averaged 4.70 (range 4.60-4.90 mm) and width 1.068 mm, head capsules averaged in length and width were 0.95 mm (range 0.90-0.100 mm) and 0.600 mm (range 0.58-0.63 mm) respectively. Mandibles averaged in length 0.065 mm (range 0.06-0.07 mm) and width 0.055 mm (range 0.052-0.058 mm). This stage lasted for 2 days.

The fifth instar larvae consumed all host tissues except alimentary canal. Tracheal system was well developed and nine pairs of spiracles were clearly seen. Two longitudinal tracheal trunks were present.

Pupa (Figure 40G)

Pupae were yellowish and exarate. In pupa two large brownish compound eyes and 3 ocelli were prominent. Appendages, such as the antennae, legs and wing pads were clear and being loosely appressed to the body. 20 pupae averaged 4.90 mm (range 4.8-5.0 mm) and width 1.28 mm (range 1.25-1.30 mm). The pupal stage lasted for 6 days.

Cocoon

The cocoons were cylindrical and grayish yellow with small back patch. 20 cocoons averaged 5.0 mm (range 4.8-5.2 mm) in length and width 1.85 mm (range 1.8-1.90 mm).

Adult (Figure 34)

The adult was brownish blakish in colour. The female measured 3.2 mm from head to tip of the abdomen. Female was recognized by upwardly curved ovipositor.

Discussion

In dispterous fly, Indian uzi fly, *Exorista bombycis,* a primary parasitoid of the mulberry silkworm *Bombyx mori* there were 3 instars (Sathe and Jadhav, 2001). Fisher (1959) observed five larval instars in *Campoletis chlorideae* Uchida (*Horogenes chrysostictus* Gmelin), a larval parasitoid of *Ephestia sericarium* Scott. Wherein the first two instars were caudate type and remaining three were hymenopteriform. While, Tikar and Thakare (1961) observed only four larval instars in the same Ichneumonid, *C. chlorideae,* an internal parasitoid of *Helicoverpa armigera* (Hubn.). Gangrade (1964) studied structure of egg, immature forms and adults of *C. chlorideae.* He reported that whitish, elongated eggs were rounded at both the ends. He further reported three moults of parasitoid larvae within the body of the host and the fourth within the cocoon before transformation into the pupal form. The first instar was with elongated tail and 13 post embryonic segments. The second instar showed reduced tail while, third instar was with great reduction of last segment. Fourth instar was comparatively large in size which occupied most of the area within the host body. Just before emergence parasitoid thrusted its head with pointed mandibles forcefully from venter of the seventh abdominal segment and came out slowly and started construction of cocoon. The present Ichneumonid, *C. charukeshi* also showed five larval instars like *C. chlorideae.* Leong and Oatman (1968) studied the life history of *Campoplex haywardi* Blanchard, a solitary internal parasitoid of tuber worm *P. operculella*, and reported only three larval instars in addition to egg, prepupa, cocoon and pupal stage. The larval stages differed in the body form, gross external appearance and mandibular size.

Oatman and Platner (1974) studied the biology of *Temelucha* sp. (Ichneumonidae), a parasitoid of potato tuber moth, *P. operculella*. Their observation deal with biometry and development of immature stages *viz.*, egg, larvae, pupa and cocoon. They also reported 3 larval instars in *Temelucha* sp. First instar was caudate with 13 body segments, second instar was soft and difficult to distinguish at its head end, while third was hymenopteriform and on thirteenth day it emerged from the host and spun its cocoon. In all present Ichneumonids, five larval instars were present, out of which first three were caudate and remaining two were hymenopteriform.

The egg of *Bathyplectes curculionis* (Thompson) (Ichneumonidae), a parasite of *Hypera postica* (Gyllenhal) nearly doubled in size during embryonation (Bertell and pass, 1978). The increase in the size of egg after deposited in host haemolymph was also observed in present forms. This was due to osmosis and active absorption of fluid through the egg membrane and this phenomenon has been reported for other endoparasitic hymenoptera (House, 1958, House *et al.*, 1971).

Campoletis sonorensis (Cameron) showed five instars and sex of the parasitoid was apparent in the fourth instar (Wilson and Ridgway, 1975). In fifth instar clear distinction of sexes was possible in *Diadegma trichoptilus* (Cameron) (Sathe, 1993). These finding are in agreement with present Ichneumonid forms. General mouth parts of all instars were similar to those described for Ichneumonid by Fisher (1959) and Wilson and Ridgway (1975) except the differences in size, shape, degree of sclerotization and some minor morphological characters.

In *Pimpla instigator* (F.) (Ichneumonidae) Rojas-Rousse and Benoit (1977) indicated five larval instars on the basis of the biochemical analysis, by measuring the mandibles and head capsules. As the instar progressed the length and width of these structures also increased relatively but it did not showed any correlation with the age of larval instars. These findings were in agreement with the general principle of moulting and growth. These observations indicate that the highly chitinized parts like mandibles and head capsules increased in size only after moulting (Rojas - Rousee and Benoit, 1974, Sathe and Margaj, 2001).

The life cycle studies of above parasitoids will be helpful for mass rearing of parasitoids and exploiting them in pest control strategies.

5

Host Age Selection

Introduction

Host age selection aspects are extremely important from view point of insect pest management. Parasitoid development rate is related with the host age, host density and longevity of parasitoids (Vinson, 1976). Parasitoids are classified into four categories by taking into account of the stage attacked *e.g.* egg, larva, pupa and adult (Doutt, 1959). All categories of parasitoids show their preferences for either younger or older developmental stages of the host (Waage and Greathead, 1986). The preference to the particular age of the host is related to hormones or alteration in the factors or alterations are necessary for their acceptance (Smilowitz, 1974).

The host age preferences of parasitoid may strongly influence the interaction between host and parasitoid and further introduction, colonization and mass production of parasitoids in planning biological control strategies. (Jowyk and Smilowitz, 1978). Host mortality rate is depends on the instars (Hogg and Nordheim 1983). Moreover, host finding rate and thus, the damage done to host plants may be affected by the instars parasitized. However, the host instar and species preferences of parasitoid have rarely been assessed (Beckage and Riddiford, 1978, Force, 1975; Van Alphen 1980; Luck *et al.,* 1982; Sathe and Margaj, 2001 etc.)

Perusal of literature indicates that host age selection in parasitoids have been studied by Sathe (1985), Chatterjee and Swarup (1961), Kumar *et al.* (1990b), Mahadevappa (1992), Sinchaisri *et al.* (1972), Sathe and Ingawale (1993), Narsimha Rao *et al.* (1993), Ram Kishore *et al.* (1993), Jyothi and Veeranna (1993), Savanurmath and Patil (1993), Thompson (1954), Prasad *et al.* (1985), Sathe and Margaj (2001) etc. The present topic was objected to find out the optimum age of pest species for maximum parasitization by parasitoids. This work will be helpful for mass rearing of parasitoids under laboratory conditions and further, field release of parasitoids at appropriate stage of the crop pest.

Materials and Methods

Rearing of pests *L. orbonalis, T. postica* and *H. armigera* were made at laboratory conditions (25 ± 1°C, 70-75 per cent R.H., 12 hr photoperiod) as per the procedure given in material and methods section of the text. The culture of pests and parasitoids were maintained in laboratory as described under material and methods chapter. Larvae of pests of known age (1 to 10 day old) were exposed to respective parasitoids for their parasitization in rearing cages (Figures 2 and 3) with a constant pest density, 30 for 24 hr. The parasitized larvae were reared in the laboratory for parasitoid emergence. After parasitism, per cent parasitism was calculated. The experiment was replicated for 5 times. During the experiments parasitoids were fed with 50 per cent honey solution and pest caterpillars on their respective host plants *viz. L. orbonalis*-brinjal fruits, *T. postica*-ber leaves and *H. armigera*-gram leaves in sufficient quantity.

Results

(1) Parasitism by *P. testaceous* in Relation to Host Age of *L. orbonalis*

30 pest larvae of each age group, 1 to 10 day old were exposed to parasitoids to determine the suitability of host age. Results are recorded in the Table 1. The parasitoids emerged from 1 - 9 day old host larvae while parasitoids have not emerged from the host age 10 days. Maximum parasitism was recorded on 3 day old host larvae and number of parasitoids

emerged under this age group was 50 (33.33 per cent). The larvae of host ranging from 2-5 day old yielded higher number of progeny. Host larvae of older than 6 days have been progressively less suitable. The sex ratio was in favour of females (1 : 1.2).

(2) Parasitism by *C. obtusus* in Relation to Host Age of *T. postica*

In each replicate 30 larvae of host with different age groups were exposed to a single mated female of *C. obtusus* to find out the most effective age for parasitization. Results are tabulated in the Table 2. Parasitoids emerged from host larvae of age 1 day to 9 day and 4 day old larvae were most suitable for parasitism, since the highest per cent of parasitization, 34 per cent with highest number of parasitoids 52 was obtained with this age group. The progeny production decreased gradually with an increase in the age of the host larvae beyond 4 days. No parasitoids were obtained from 10 day old host larvae. The sex ratio was in favour of females 1 : 1.04.

(3) Parasitism by *D. fenestralis* in Relation to Host Age of *H. armigera*

The larvae of *H. armigera* of known age groups ranging from 1 to 10 day were exposed to newly mated females of *D. fenestralis* in oviposition unit for 24 hr to determine the most suitable age for parasitism. 3 day old host larvae produced highest number of adult parasitoids 54 (36 per cent). The progeny production decreased gradually with an increase in the age of the host larvae beyond 3 day. The host larvae of age 1 day to 9 day old were susceptible for progeny production. The host larvae 10 day old remain unparasitized. The sex ratio was favouring the females, 1 : 1.13 (Table 3).

Discussion

The pest preferences of parasitoids to immature stages have been studied by several workers (Miller, 1970; Leong and Oatman, 1968; Sathe and Margaj, 2001 etc.). The uzi flies of silkworm have been attempted by several workers (Sinchaisri *et al.,* 1972; Prasad *et al.,* 1985; Kumar *et al.,* 1990; Narsimha Rao *et al.,* 1993; Ram Kishore *et al.,* 1993; Savanurmath and Patil, 1993 etc.) with respect to host age selection. Richerson and Deloach (1972) says that the host size can change the host choice by parasitoid. Gutierrez (1970) opines that age dependent size of host was

Table 1: Maximum Effective Age of *L. orbonalis* Larvae for Parasitism by *P. testaceous*.

Host Age (Days)	Total Larvae Exposed	No. of Moths Emerged	Total No. of Parasitoids Emerged				Parasitism per cent	Pest Mortality (Unknown)
			Male	Female	Total	Sex Ratio (M : F)		
1	150	133	6	5	11	1 : 0.83	7.33	4
2	150	122	12	13	25	1 : 1.08	16.66	3
3	150	100	24	26	50	1 : 1.08	33.33	0
4	150	110	16	23	39	1 : 1.43	26.00	1
5	150	118	14	17	31	1 : 1.21	20.60	1
6	150	128	9	12	21	1 : 1.33	14.00	1
7	150	133	7	8	15	1 : 1.14	10.00	2
8	150	137	5	5	10	1 : 1.00	6.66	3
9	150	142	3	3	6	1 : 1.00	4.00	2
10	150	145	0	0	0	0 : 0	0.00	5
					Avg.	1 : 1.2		

Table 2: Maximum Effective Age of *T. postica* Larvae for Parasitism by *C. obtusus.*

Host Age (Days)	Total Larvae Exposed	No. of Moths Emerged	Total No. of Parasitoids Emerged				Parasitism per cent	Pest Mortality (Unknown)
			Male	Female	Total	Sex Ratio (M : F)		
1	150	135	4	4	8	1 : 1.00	5.33	7
2	150	122	15	9	24	1 : 0.6	16.00	4
3	150	110	18	18	36	1 : 1.00	24.00	4
4	150	98	24	28	52	1 : 1.16	34.00	0
5	150	120	20	28	48	1 : 1.40	32.00	2
6	150	128	8	13	21	1 : 1.62	14.00	1
7	150	133	8	8	16	1 : 1.00	10.66	1
8	150	137	6	7	13	1 : 1.16	8.66	0
9	150	142	2	2	4	1 : 1.00	2.66	2
10	150	146	0	0	0	0 : 0	0.00	4
					Avg.	1:1.04		

Table 3: Maximum Effective age of *H. armigera* Larvae for Parasitism by *D.fenestralis.*

Host Age (Days)	Total Larvae Exposed	No. of Moths Emerged	Total No. of Parasitoids Emerged				Parasitism per cent	Pest Mortality (Unknown)
			Male	Female	Total	Sex Ratio (M : F)		
1	150	133	7	6	13	1 : 0.85	8.66	4
2	150	104	20	22	42	1 : 1.10	28.00	4
3	150	95	25	29	54	1 : 1.16	36.00	1
4	150	108	20	21	41	1 : 1.05	21.33	1
5	150	110	16	18	34	1 : 1.12	22.66	6
6	150	120	15	15	30	1 : 1.00	20.00	0
7	150	133	10	12	22	1 : 1.20	14.66	5
8	150	137	7	5	12	1 : 0.71	8.00	1
9	150	142	2	4	6	1 : 2.00	4.00	2
10	150	145	0	0	0	0 : 0	0.00	5
						Avg. 1:1.13		

also responsible to influence the host acceptance. Leong and Oatman (1968) says that there is relationship between host age and rate of parasitization in parasitoid host model.

Sinchaisri *et al.* (1972) have studied the host preferences of the Tachinid fly to different stages and varieties of silkworms and they concluded that *Tachina sorbillans* prefer to lay their eggs on the later stages and Japanese varieties to younger stages of silkworm, probably due to larger body size. The observations made by Narsimha Rao *et al.* (1993) on *Exorista sorbillans* Wiedemann a parasitoid of *Bombyx mori* L. reveals that uzi fly parasitized the developmental stages of silkworm including chawki worms, except 1st instar. The infestation of the fly was severe with the advancement of development (Chawki to 5th instar) confirming the view of Siddappaji and Basavanna (1990) that ovipositing flies preferred older larvae. These observations also lead a strong support to the suggestions of Kasturibai *et al.* (1986) that the 4th and 5th instar silkworms produce a kairomone which is strongly attractive to the uzi fly. The instar wise percent infestation was 6.59, 6.16, 9.35 and 16.03 per cent for 2nd, 3rd, 4th and 5th instars respectively. Jyothi and Veeranna (1993) reported that in *E. sorbillans,* the host pupae exposed for parasitization at ages ranging 4-5 day old pupae were suitable for infestation.

In *Apanteles* (= *Pseudoapanteles*) *dignus* (Meusebeck) (Braconidae), a parasitoid of *Keiferia iycopersicella* (Walsingham) maximum parasitism was recorded on 2-3 day old host larvae and the higher number of parasitoids per replicate was 12 individuals. Decrease in the progeny production was observed gradually with an increase in the age of the host larvae beyond 2-3 day old. No parasitoids were obtained when larvae of 8-9 day old were exposed (Cardona and Oatman, 1971).

Oatman and Platner (1974) says that 3-4 day old caterpillars of *Phthorimaea opercullella* (Zeller) were most suitable for maximum parasitization by *Temelucha* sp. of *Platensis* group (Ichneumonidae). They also noted that *Hyposoter exoguae* (Viereck), a solitary endoparasitoid of the cabbage looper, *Trichoplusiani* (Hubner) preferred to oviposit in young host larvae and showed greatest preference for 1st instars (Smilowitz and Iwanstsch, 1975) by maximum of 31 per cent parasitism. However, decline in parasitism was noted in further instars.

Some parasitoids kill their hosts by feeding on them and oviposit in others. The Apheinid, *Encarsia formosa* oviposit in white fly larvae of the 3rd and 4th instar and prepupae but, feeds only on 2nd instars and pupae (van Alphen *et al.*, 1976). Van Alphen (1980) found that the Eulophid *Tetrastichus asparagi* Crawford used young eggs of host for feeding but oviposited in eggs with a developed embryo. However, some Ichneumonid parasitoids like *C. chlorideae*, a parasitoid of Gram pod borer, *Helicoverpa armigera* (Hubn.) and *Eriborus argenteopilosa* (Cameron) feed on host haemolymph oozed from the host at the time of oviposition. (Sathe and Santhakumar, 1988a; Sathe, 1990b). In *Tetrastichus asparagi* Crawford handling time of older hosts was much longer because of defense by the host.

According to Thurston and Postley (1978), the development rate of a parasitoid is dependent on the host instar parasitized. These workers worked with *Apanteles* (=*Cotesia*) *congregatus* (Say), a parasitoid of *Manduca sexta* (L.) and observed that the development of the parasitoid was slower when oviposition occurred in 1st or 2nd instars than occurred in 3rd or 4th. Leibee *et al.* (1979) reported optimum parasitization by *Potsson lameeres* Debauch (Hymenoptera : Mymaridae) on 2 day old *Sitona hispidulus* (Fabricius) larvae. They further reported that females have difficulty in inserting their ovipositor into older eggs. Therefore, it was suggested that the reduced parasitism after the optimum age was due to the harder chorion of the older eggs. The above phenomenon was not observed in present parasitoids since they are larval parasitoids.

According to Sathe and Margaj (2001) the solitary parasitoid prefers younger hosts whereas, the gregarious older ones. The reason in solitary parasitoids in doing so would be generally one egg/eggs, are/is laid in a single host and only one parasitoid develops among them, ii) the nutrient requirement for the development of single parasitoid is less, iii) while depositing their eggs, the females have to face the resistance exhibited by older hosts, as the host become more defendant. While in gregarious forms, i) number of parasitoids have to develop from a single host, ii) the parasitoids require relatively large quantity of food, iii) probably the hosts might be less defendant.

Several species of parasitic wasps discriminate hosts of different size for development of appropriate sex (Kishi, 1970; van den Assem *et al.*,

1984; Sandlan, 1979; Jones *et al.,* 1982; van den Assem *et al.,* 1984). The small hosts were prefereably used for male offspring and large for female (Charnov *et al.,* 1981) but, in *Lariophagus distinguendus,* a solitary larval parasitoid of *Sitophilus grannarium* very few males emerged from larger hosts and hardly any male from larger and intermediate size host, indicating that the host size was an important factor in the chain of process which lead to sex determination (Simboloti *et al.,* 1987). In the present Ichneumonids, the sex ratio in older instars was favouring the females. The findings are in agreement with the opinion of Charnov *et al.* (1981).

Lingren *et al.* (1970) studied the host age preference by *C. chlorideae* (Ichneumonidea) towards four lepidopterous host species *viz. Pseudoletia unipuncta* (Howoth), *Trichoplusiani, Prodenia eridian* (Cramer) and *Pseudoletia praefica* Grote. They observed that larvae of 1-8 day old of all the hosts were susceptible for parasitism, 2-6 days old were readily parasitized and 2-4 day old being most acceptable. In *C. charukeshi* 1-10 day old larvae were susceptible for parasitism, 2-6 day old were readily parasitized and maximum parasitization was observed on 3 day old larvae. In general, it is concluded that, host age selection by parasitoid has tremendous importance since host selection by parasitoid tell us exact stage of the pest attached by the parasitoid and exact time (stage specific) of release of parasitoids for biological control of the pest.

6

Host Density

Introduction

In mass rearing, release and colonization of parasitoids host density plays a very crucial role. Host density has relation with parasitism both under laboratory and field conditions (Sathe, 2004). In fact, biocontrol programme success is dependent on host density factors. However, very little is known on host parasitoid density dependant factors (Sathe and Margaj, 2001).

Review of literature indicates that Thompson (1922, 1924), Nicholson and Bairlay (1935), Hassell and Varley (1969), Khan and Verma (1945), Leong and Oatman (1968), Oatman *et al.* (1969), Yeargan and Latheef (1976), Sathe (1985, 1990) etc. attempted the work releated to host density and parasitism. Host density for maximum parasitism was studied in following Ichneumonids, *P. testaceous* on *L. orboralis*, *C. obtusus* on *T. postica* and *D. fenestralis* on *H. armigera*.

Materials and Methods

Cultures of parasitoids and their respective hosts were maintained under laboratory conditions (25 ± 1°C, 70 - 75 per cent R.H., 12 hr photoperiod). 3-4 day old larvae of *C. orbonalis,* 4-5 day old larvae of *T. postica* and 3 day old larvae of *H. armigera* were exposed in densities. 10, 20, 40, 50 and 100 to mated females of *P. testaceous, C. obtusus* and *D. fenestralis* respectively for 24 hr in an oviposition cage (Figure 2). After exposure the host larvae were separated into plastic containers for further observations. During the experimental work the parasitoids were fed with 50 per cent honey solution and the host larvae with their respective food plants as given in materials and methods. Observations were taken at 12 hr intervals and experiments were replicated for five times.

Results

(1) Parasitism by *P. testaceous* in Relation to Host Density of *L. orbonalis*

Results recorded in Table 1 indicated that the number of parasitoids obtained from host density 30 was highest, 51 compared to those produced from other host densities 10, 20, 30 and 100. The mean percentage of parasitism was highest 34.00 with 30 host density, whereas host densities 10, 20, 50 and 100 showed 28, 31, 30 and 30.80 mean percentage of parasitism respectively. Hence, host density 30 is to be considered as optimum, required for production of maximum parasitoid progeny.

(2) Parasitism by *C. obtusus* in Relation to Host Density of *T. postica*

The results recorded in Table 2 indicates that the number of parasitoids at host density 50 was the highest compared to those produced from other host densities. The mean percentage of parasitism, 34.00 per cent was also the highest at this density. It was found that 26.00, 28.00, 33.5 and 32.4 per cent parasitoids have emerged from host densities 10, 20, 40 and 100 respectively.

(3) Parasitism by *D. fenestralis* in Relation to Host Density of *H. armigera*

The results are recorded in Table 3. The mean percentage of parasitism was highest 36.50 with 40 host density, where as host densities, 10, 20, 50 and 100 have 30.00, 32.00, 34.80 and 33.20 mean percentage of parasitism respectively. Hence, host density 40 can be considered as optimum, for production of maximum parasitoid progeny of *D. fenestralis.*

Table 1: Effect of Host Density on Parasitism by *P. testaceous*.

Replicate No.	Host Density	No. of Host Larvae Exposed	Total No. of Parasitoids Obtained			Percentage of Emergence of Months	Mean Percentage of Parasitism
			Male	Female	Both		
1	10	50	7	7	14	72.00	28.00
2	20	100	13	18	31	69.00	31.00
3	30	150	25	26	51	64.67	34.00
4	50	250	30	45	75	70.00	30.00
5	100	500	76	78	154	69.20	30.80

Table 2: Effect of Host Density on Parasitism by *C. obtusus*.

Replicate No.	Host Density	No. of Host Larvae Exposed	Total No. of Parasitoids Obtained			Percentage of Emergence of Months	Mean Percentage of Parasitism
			Male	Female	Both		
1	10	50	7	6	13	74.00	26.00
2	20	100	12	16	28	72.00	28.00
3	40	200	28	39	67	66.50	33.50
4	50	250	41	43	85	66.00	34.00
5	100	500	80	82	162	67.60	32.40

Table 3: Effect of Host Density on Parasitism by *D. fenestralis.*

Replicate No.	Host Density	No. of Host Larvae Exposed	Total No. of Parasitoids Obtained			Percentage of Emergence of Months	Mean Percentage of Parasitism
			Male	Female	Both		
1	10	50	7	8	15	70.00	30.00
2	20	100	14	18	32	68.00	32.00
3	40	200	35	38	73	63.50	36.50
4	50	250	41	46	87	65.20	34.80
5	100	500	79	87	166	66.80	33.20

Discussion

According to Leong and Oatman (1968) in *C. haywardi* an Ichneumonid parasitoid of *Phthorimaea operculella* (Zeller), the optimum host density was 75 larvae per tuber. There was great variation between replicates at the same larval densities. The mean emergence at higher densities was from 3 per cent to 15 per cent less than at 75 host density. Cardona and Oatman (1971) reported 90 optimum host density of *Keiferia lycopersicella* (Walsingham) for maximum parasitism by *Pseudapanteles dignus* (=*A. dignus*). Oatman and Platner (1974) studied density dependent relationships in *Temelucha* sp. an Ichneumonid parasitoid of potato tuber moth *P. operculella*. Their observations were also in conformity with those of Oatman *et al.* (1969) and they found that single tuber infected with 150 larvae of 3-4 day old have proved to be optimum.

Yeargan and Latheef (1976) reported host parasitoid relationships with different densities between *Hypera postica* (Gyllenhal) and its Ichneumonid parasitoid, *Bathyplectis curculionis.* They reported uneven host densities was the only one component of its overall responses to host abundance which does not rule out the possibility of a dependant relationship between *B. curculionis* and *H. postica* and indicated that parasitoid may have a random searching pattern with regard to host density.

Sathe (1985) studied influence of host density on percentage parasitism by *D. trichoptilus*, a larval parasitoid of *Exelastis atomosa* and reported that highest number of parasitoids obtained when 30 larvae were exposed for oviposition. At this density, the maximum parasitism was 21.66 per cent while the percentage of parasitism with 10, 20, 40 and 50 densities was 12.50, 17.50, 15.93 and 12.50 respectively.

Sathe (1990) also studied effect of host density on parasitization by *D. argenteopilosa* (Hymenoptera : Ichneumonidae), a larval parasitoid of *Spodoptera litura* (Fab.) wherein he reported optimum 50 host density for maximum parasitization and progeny production.

Sathe and Bhosale (2011) studied host density relationship in *Diadegma insularae* (parasitoid) and *Plutella xylostella* (host). They recorded the highest percentage of parasitism (30.50) with 100 host density. On 5, 25, 50, 150 and 200 host densities maximum progeny production per cent were 22.00, 37.00, 38.00, 37.50 and 38 respectively.

In the present study optimum host densities for maximum progeny production were found to be 30, 50 and 40 for *P. testaceous, C. obtusus* and *D. fenestralis* respectively. This work will be helpful for biological control of insect pests mentioned in the text.

7

Fecundity and
Intrinsic Rate of Increase

Introduction

As a part of ecofriendly approach of pest management, data are required on several basic population processes, ovipositional rate, developmental rate, intrinsic rate etc. (Ruesink, 1976). Therefore, the primary objectives of the present topic were to find out the longevity, ovipositional period, progeny production and sex ratio for assessment of reproductive potentials of Ichneumonid parasitoids as biocontrol agents of insect pests and second objective was to investigate intrinsic rate of natural increase in Ichneumonid parasitoids.

Fecundity and intrinsic rate of Ichneumonid parasitoids have been studied by Leong and Oatman (1968) in *Campoplex haywardi* Blanchard, Oatman and Platner (1974) in *Temelucha* sp. *Platensis* group (Ichneumonidae), Chundurwar (1975) in *Eriborus trochanteratus* Morley, Basarkar and Nikam (1981) in *Xanthopimpla stemmator* (Thunberg), Yeargan and Latheef (1977) in *Bathyplectes anurus* Thompson, Yeargan (1979) in *Bathyplectes stenostigma* Thompson, Basarkar and Nikam (1981) in

Goryphus nursei (Cameron) (Ichneumonidae), Sathe and Nikam (1986) in *Campoletis chlorideae* (Uchida) and Sathe (1987) in *Diadegma trichoptilus*.

Fecundity is helpful to determine intrinsic rate of natural increase in natural population. Life table studies, fecundity and survivorship combined with statistical determination enables to evaluate the population growth or population decline parameters of an organism with respect to various environmental factors affecting population (Birch, 1948).

Ichneumonids are pest population regulatory factors hence investigators are being attracted to investigate the mechanisms of population regulation both with species and biosenoses (Thompson, 1924). Thompson (1924) for the first time developed mathematical method for studying the population dynamics of insects. Later, Lotka (1925) developed a function to count the birth rate and death rate with respect to age. Birch (1948) for the first time used the function of intrinsic rate of increase for the insect population.

The intrinsic rate of natural increase is the actual rate of increase of population under specialized constant environmental conditions where space and food are unlimited and no mortality factors are present other than physiological (Birch, 1984) and calculated by the following formula:

$$\Sigma = e\ 7^{-r}\ m \times l_x\ m_x = 1$$

where, 'e' is the base of natural logarithms, 'x' the age of the individual in days, l_x the number of individual alive at age 'x' as a proportion of one, 'm_x' the number of female offsprings produced per female in the age interval 'x'. The sum of the products $l_x m_x$ is the net reproductive rate, 'R_o' is the rate of multiplication of the population in each generation measured in terms of female produced per generation.

The approximate value of cohort generation time 'T_c' was calculated as:

$$T_c = \frac{l_x m_x\ x}{l_x m_x}$$

The arbitrary value of innate capacity of increase 'r_c' was calculated by following function

$$r_c = \frac{\log_e R_o}{T_c}$$

This was arbitrary value for 'r_m' and value of 'r_m' up to two decimal places was substituted in the formula until the two values of the equation were found which lies immediately above or below 1096.6.

$$\Sigma = e\ 7^{-r}\ m \times l_x\ m_x = 1$$

The point of intersection provides the value of 'r_m' accurate, the 3 decimal places. The precise generation time 'T' was then calculated by

$$T = \frac{\log_e R_o}{r_m}$$

and the finite rate of increase (λ) was calculated as $\lambda = e^r m$.

Materials and Methods

Under laboratory conditions (25±1°C, 70-75 per cent R.H., 12 hr. photoperiod). Cultures of parasitoids and their hosts were maintained and used in the experiments. For fecundity studies the hosts of age 3, 4 and 3 day old were exposed to the parasitoids, *P. testaceous, C. obtucus* and *D. fenestralis* respectively with constant 50 host density for 24 hr till the death of females. The parasitized hosts were reared in separate containers. The observations were made on immature forms, longevity of adult parasitoids, life cycle, daily parasitoid and moth emergence and sex ratio from each lot of hosts. The life tables were constructed with the help of fecundity data and later 'r_m's were calculated in parasitoids.

Results

(1) Fecundity and Intrinsic Rate of Increase in *P. testaceous*

Longevity of ovipositing females averaged 12.6 days (range 10-14 days). The male : female offsprings averaged 1 : 1.09 (range 1 : 1 - 1 : 1.40). Most of the females reached their peak of oviposition on the fourth day. Average number of progeny obtained was 34.3 (range 32-38) individuals. Female oviposition period averaged for 8.3 days (range 7-9 days). The first adult mortality was on the tenth day. Average length of immature stages of parasitoid was 22 days. The maximum progeny production per day, m_x was 41 on fourth day and reproduction stopped on 9th day. The intrinsic rate of increase per female per day was 0.107 and population multiplied 53.4 times in mean generation time 'T' of 18.58 days. Results are recorded in Tables 1, 2 and 3.

Table 1: Longevity, Oviposition, Fecundity and Sex Ratio of Mated Females *P. testaceous*.

Replicates	Longevity (Days)	Oviposition Period (Days)	No. of Larvae Exposed	Parasitoid Progeny			Sex Ratio Male : Female
				Male	Female	Total	
A	12	8	50	16	17	33	1 : 1.06
B	11	8	50	15	21	36	1 : 1.40
C	14	9	50	15	18	33	1 : 1.20
D	14	9	50	17	18	35	1 : 1.05
E	14	9	50	17	17	34	1 : 1.0
F	12	8	50	17	18	35	1 : 1.05
G	13	8	50	16	17	33	1 : 1.06
H	10	7	50	16	16	32	1 : 1.0
I	14	9	50	18	20	38	1 : 1.11
J	12	8	50	17	17	34	1 : 1.00
Avg.	12.6	8.3	50.0	16.4	17.9	34.3	1 : 1.09

Table 2: Daily Progeny Production of Mated Females of *P. testaceous*.

Replicates	Number of Female Progeny Production per Day per Female															
	1	2	3	4	5	6	7	8	9	10	11	12	13	14	15	Total
A	0	1	2	4	4	3	2	1	0	D	–	–	–	–	–	17
B	1	2	3	5	4	3	2	1	0	0	D	–	–	–	–	21
C	0	1	2	3	4	3	2	2	1	0	0	0	0	D	–	18
D	0	1	2	3	5	3	2	1	1	0	0	0	0	D	–	18
E	0	1	2	3	3	3	2	2	1	0	0	0	0	D	–	17
F	1	2	4	4	2	2	2	1	0	D	–	–	–	–	–	18
G	0	1	2	5	3	3	1	1	0	0	0	0	D	–	–	17
H	1	2	2	5	3	2	1	0	0	D	–	–	–	–	–	16
I	1	2	3	4	3	3	2	1	1	0	0	0	0	D	–	20
J	1	1	2	5	3	2	2	1	0	0	0	D	–	–	–	17
Avg.	0.5	1.4	2.4	4.1	3.4	2.7	1.9	1.1	0.4	0.0	0.0	0.0	0.0	0.0	0.0	

Table 3: Life Table Statistics of *P. testaceous.*

Pivotal Age (Days) x	Proportional Live at Age l_x	No. of Female Progeny/Female m_x	$l_x m_x$	$l_x m_x x$
Immature stages - 22 days				
23	1	0.5	0.5	11.5
24	1	1.4	1.4	33.60
25	1	2.4	2.4	60.00
26	1	4.1	4.1	106.60
27	1	3.4	3.4	91.80
28	1	2.7	2.7	75.60
29	1	1.9	1.9	55.10
30	1	1.1	1.1	33.00
31	0.6	0.4	0.24	0.7.44
32	0.0	0.0	0.0	0.00
		Avg.	17.74	474.64

$$Tc = \frac{l_x m_x \; x}{L_x m_x} = \frac{474.64}{17.74} = 26.75$$

$$r_c = \frac{\log_e R_o}{T_c} = \frac{\log_e 17.74}{26.75} = 0.107$$

r_c is arbitrary r_m

$r_m = 0.107$

(2) Fecundity and Intrinsic Rate of Increase in *C. obtusus*

Results are recorded in Tables 4 to 6. Longevity of ovipositing female averaged 13.5 days (range 11-16 days) and progeny production 43 (range 41-48) individuals. The male : female ratio was 1 : 1.19 (range 1 : -1 : 1.5). Most of the females reached their peak of oviposition on the 5th day. The first adult mortality was on the 11th day. Oviposition period averaged 10.7 days (ranged 8-13 days). Average length of immature stages of parasitoid was 20 days. The maximum progeny production per day, m_x was 38.00 on 5th day and reproduction stopped on 14th day. The intrinsic rate of increase per female per day was 0.119 and population multiplied 22.8 times in mean generation time 'T' of 17.87 days.

Table 4: Longevity, Oviposition, Fecundity and Sex Ratio of Mated Females *C. obtusus.*

Replicates	Longevity (Days)	Oviposition Period (Days)	No. of Larvae Exposed	Parasitoid Progeny			Sex Ratio Male : Female
				Male	Female	Total	
A	11	8	50	20	21	41	1 : 1.05
B	13	10	50	20	22	42	1 : 1.10
C	16	10	50	22	26	48	1 : 1.18
D	16	13	50	20	26	46	1 : 1.30
E	14	12	50	21	24	45	1 : 1.14
F	14	12	50	19	23	42	1 : 1.21
G	15	12	50	20	22	42	1 : 1.10
H	13	12	50	18	23	41	1 : 1.27
I	11	8	50	21	21	42	1 : 1.0
J	12	10	50	20	23	43	1 : 1.50
Avg.	13.5	10.7	50.00	20.1	23.1	43.2	1 : 1.19

Table 5: Daily Progeny Production of Mated Females of *C. obtusus*.

Replicates	Number of Female Progeny Production per Day per Female																
	1	2	3	4	5	6	7	8	9	10	11	12	13	14	15	16	Total
A	0	1	2	3	6	3	4	2	0	0	D	—	—	—	—	—	21
B	0	1	2	2	5	5	3	2	1	1	0	0	D	—	—	—	22
C	1	1	1	3	5	4	5	3	2	1	0	0	0	0	O	D	26
D	0	2	1	3	3	3	4	3	2	2	1	1	1	0	0	D	26
E	0	1	1	2	4	3	4	3	2	2	1	1	0	D	—	—	24
F	0	1	1	2	2	3	3	3	3	2	2	1	O	D	—	—	23
G	0	1	1	2	2	2	3	2	3	2	2	2	0	0	D	—	22
H	0	1	2	2	2	3	3	3	2	2	2	1	D	—	—	—	23
I	1	2	3	3	5	4	2	1	0	0	D	—	—	—	—	—	21
J	1	2	3	4	4	3	2	2	1	1	0	D	—	—	—	—	23
Avg.	0.3	1.3	1.7	2.6	3.8	3.3	3.3	2.4	1.6	1.3	0.8	0.6	0.1	0	0	0	

Table 6: Life Table Statistics of *C. obtusus.*

Pivotal Age (Days) x	Proportional Live at Age l_x	No. of Female Progeny/Female m_x	$l_x m_x$	$l_x m_x x$
Immature stages - 20 days				
21	1	0.3	0.3	6.30
22	1	1.3	1.3	28.60
23	1	1.7	1.7	39.10
24	1	2.6	2.6	62.40
25	1	3.8	3.8	95.00
26	1	3.3	3.3	85.80
27	1	3.3	3.3	89.10
28	1	2.4	2.4	67.2
29	1	1.6	1.6	46.40
30	1	1.3	1.3	39.00
31	0.80	0.80	0.64	19.84
32	0.60	0.60	0.36	11.52
33	0.40	0.10	0.04	1.32
34	0.20	0.00	0.00	0.00
		Avg.	22.80	591.58

$$Tc = \frac{l_x m_x\, x}{L_x m_x} = \frac{591.58}{22.84} = 25.94$$

$$r_c = \frac{\log_e R_0}{T_c} = \frac{\log_e 22.80}{25.94} = 0.120$$

r_c is arbitrary r_m

$r_m = 0.120$

(3) Fecundity and Intrinsic Rate of Increase in *D. fenestralis*

Results are recorded in Tables 7 to 9. Longevity of ovipositing females averaged 18.50 days (range 15 to 20 days). The number of progeny produced averaged 54.5 individuals (range 43 to 62 individuals). The male : female offsprings averaged 1 : 1.26 (range 1 : 1.04 to 1 : 1.30). Progeny production peak of female was on 6th day, the first adult mortality was on

Table 7: Longevity, Oviposition, Fecundity and Sex Ratio of Mated Females *D. fenestralis.*

Replicates	Longevity (Days)	Oviposition Period (Days)	No. of Larvae Exposed	Parasitoid Progeny			Sex Ratio Male : Female
				Male	Female	Total	
A	18	12	50	25	29	54	1 : 1.26
B	18	13	50	22	25	47	1 : 1.13
C	20	15	50	20	26	46	1 : 1.30
D	20	15	50	25	31	56	1 : 1.24
E	19	14	50	21	26	47	1 : 1.23
F	19	13	50	20	23	43	1 : 1.15
G	20	15	50	21	23	44	1 : 1.09
H	18	13	50	24	25	49	1 : 1.04
I	15	13	50	23	24	47	1 : 1.04
J	18	14	50	24	26	50	1 : 1.08
Avg.	18.50	13.7	50.00	25.5	29.0	54.5	1 : 1.26

Table 8: Daily Progeny Production of Mated Females of *D. fenestralis*.

Replicates	Number of Female Progeny Production per Day per Female																Total
	1	2	3	4	5	6	7	8	9	10	11	12	13	14	15	16	
A	0	1	2	2	4	5	5	4	3	2	1	0	0	D	—	—	29
B	0	2	2	2	3	6	4	3	2	1	0	D	—	—	—	—	25
C	0	1	2	3	4	6	4	2	2	1	1	0	D	—	—	—	26
D	1	1	2	4	5	5	6	3	2	1	1	0	D	—	—	—	31
E	0	0	2	5	5	6	3	2	2	1	0	D	—	—	—	—	26
F	0	1	2	3	4	6	3	2	1	1	D	—	—	—	—	—	23
G	0	2	2	2	5	6	4	2	1	1	0	D	—	—	—	—	25
H	1	1	3	3	6	4	3	2	1	0	D	—	—	—	—	—	24
I	0	1	2	3	5	5	4	3	2	1	0	D	—	—	—	—	26
J	0	1	2	3	4	5	5	4	3	2	1	0	D	—	—	—	32
Avg.	0.2	1.1	2.1	3.0	3.5	5.4	4.1	2.7	1.8	1.1	0.4	0	0	0	0	0	

Table 9: Life Table Statistics of *D. fenestralis.*

Pivotal Age (Days) x	Proportional Live at Age l_x	No. of Female Progeny/Female m_x	$l_x m_x$	$l_x m_x x$
Immature stages - 16 days				
17	1.00	0.2	0.2	3.40
18	1.00	1.1	1.10	19.80
19	1.00	2.1	2.10	39.90
20	1.00	3.0	3.00	60.00
21	1.00	3.5	3.50	73.50
22	1.00	5.4	5.40	118.80
23	1.00	4.1	4.10	94.30
24	1.00	2.7	2.7	64.80
25	1.00	1.8	1.8	45.00
26	1.00	1.1	1.1	28.60
27	0.80	0.4	0.32	8.64
28	0.40	0.0	0.0	0.0
29	0.30	0.0	0.0	0.0
30	0.10	0.0	0.0	0.0
31	0.0	0.0	0.0	0.0
		Avg.	25.32	556.74

$$Tc = \frac{l_x m_x\, x}{L_x m_x} = \frac{556.76}{25.32} = 21.98$$

$$r_c = \frac{\log_e R_0}{T_c} = \frac{\log_e 25.32}{21.98} = 0.147$$

r_c is arbitrary r_m

$r_m = 0.147$

the 10th day and oviposition period averaged 13.7 days with the range 12 to 15 days. Average length of immature stages of parasitoid was 16 days. The maximum progeny production per day, m_x (54.00) was on 6th day and reproduction stopped on 12th day. The intrinsic rate of natural increase per female per day was 0.147 and population multiplied to 25.32 times in mean generation time 21.98 days.

Discussion

In *Campolex haywardi* Blanchard progeny production ranged from 44 to 143 (average 87.3), longevity varied from 4 to 19 days (average 10.5 days) and male : female ratio of offsprings was between 1.0 : 0.06 to 1.0 : 1.93 (average 1.0 : 0.65) (Leong and Oatman, 1968). The progeny production ranged from 40 to 51 (average 44.6) in *Diadegma trichoptilus* (Cameron) (Sathe, 1982).

Broodryk (1969b) studied the fecundity and sex ratio of a braconid *O. parcus* by providing abundant hosts and observed that parasitoid progeny per female was 64.8 per cent, the maximum and minimum being 127 and 64.8 respectively with sex ratio of 2.2 male : 1 female. The fecundities were very high in *O. lepidus* (Braconidae) (Oatman *et al.*, 1969) and in *Temelucha* sp. (Ichneumonidae) (Oatman and Platner, 1974) as compared to the fecundities of species studied by Oatman and Platner (1974) and this may be due to the solitary nature of the parasitoids. Sex ratio did not showed significant differences in the present forms and the other species studied previously.

In an Ichneumonid, *Eriborus trochanteratus* (Morley) the life table and intrinsic rate of increase was studied by Chundurwar (1975) and noted that the intrinsic rate of increase was 0.160 and the population multiplied to 30.56 times in mean generation time of 19.10 days.

In another Ichneumonid *Xanthopimpla stemmator* (Thunburg), a pupal parasitoid of *Chilo partellus,* the intrinsic rate of natural increase was 0.131 while, population multiplied to 43.43 times in 28.78 days of mean generation time (Basarkar and Nikam, 1981). Sathe (1990) studied longevity, oviposition, sex ratio and fecundity in *Diadegma argenteopilosa* Cameron (Hymenoptera : Ichneumonidae), a parasitoid of *Spodoptera litura* (Fab.). He reported 15.1 days as average longevity, 9 days as ovipositional period and 62.00 adults as average progeny production with sex ratio (M:F) 1 : 1.071.

In *Diadegma trichoptilus* (Cameron) intrinsic rate of increase was 0.149 and population multiplied to 25.63 times in mean generation time 21.77 days. Immature duration reported was 18 days in this parasitoid (Sathe, 1988). Sathe and Nikam (1986) also studied life tables and intrinsic rate of increase in *Campoletis chlorideae* (Uchida), a parasitoid of *H. armigera* wherein the intrinsic rate of increase was 0.112, average oviposition period was 9.6 days and average progeny production was 48.17 individuals. The average longevity of ovipositing females was 11.2 days with sex ratio (m:f) 1 : 2.06. Population multiplied to 15.24 times in the mean generation time, 'T' 24.31 days while, in the present parasitoid, *D. fenestralis* intrinsic rate of increase was 0.147 with sex ratio (m:f) 1:1.26 individuals, the average longevity of ovipositing females was 12 to 15 days with sex ratio (m:f) 1 : 1.26 (range 1 : 1.04 to 1 : 1.30) and population multiplied 25.32 times in mean generation time 'T' of 21.98 days. In the present study the intrinsic rates of increase were 0.107, 0.119 and 0.149 in *P. testaceous, C. obtusus* and *D. fenestralis* respectively.

8

Ecobiology: Effect of Temperature on Development and Survival

Introduction

Temperature affects development, longevity, fecundity, sex-ratio, survival and parasitism by parasitoid (Mellini *et al.,* 1979; Nikam and Sathe, 1981). For mass rearing of parasitoids optimum temperature is crucial factor. Therefore, development of Ichneumonids (Hymenoptera : Ichneumonidae) as a biocontrol agents of pests has been studied at different temperatures. Review of literature indicates that Beling (1933), Fisher (1959), Nikam and Basarkar (1978), Nikarm and Sathe (1981) etc. attempted the studies related to temperature effect on development of Ichneumonid parasitoids.

Materials and Methods

Three to four day old larvae of pests were exposed to newly emerged and mated females of Ichneumonid parasitoids namely *P. testaceous, C. obtusus* and *D. fenestralis* in breeding glass cages (size 25 × 25 × 30 cm) for 6 hr for parasitism. Parasitized larvae were transferred into separate containers (size 6.5 × 5.00 cm) for further rearing. Host food plant parts were provided to host caterpillars as food during the rearing and 50 per cent honey solution to parasitoids. Effect of temperature on the development and survival of the parasitoid has been studied at 5±1°C, 10±1°C, 15±1°C, 20±1°C, 25±1°C, 30±1°C, 35±1°C, and 40±1°C, in BOD incubator and seed germinator. Observations on the development of the parasitoid were made at 12 hr interval. Experiments were replicated for five times and in each lot 10 individuals were taken into account.

Results

P. testaceous

The results are recorded in Table 1. At 5±1°C, 10±1°C, 35±1°C, and 40±1°C the development was not possible. However, at 35±1°C pupal development was possible. The eggs hatched within 5.00, 4.50, 3.00 and 3.00 days; the larval period was 13.00, 12.50, 12 and 11 days and pupal period was 7.00, 6.00, 6.00 and 5.00 days; thus the life cycle completed within 25, 23, 21 and 19 days at 15±1°C, 20±1°C, 25±1°C, and 30±1°C respectively. The optimum temperature for development of parasitoid was 25±1°C and the average percentage of survival was 77.00, 80.00, 90.00 per cent and 85.00 per cent at 15±1°C, 20±1°C, 25±1°C and 30±1°C respectively.

Table 1: Effect of Temperature on Development of *P. testaceous.*

Temperature (±1°C)	Incubation Period (Days)	Larval Period (Days)	Pupal Period (Days)	Total Period for Development (Days)	Per cent Survival
5	0.00	0.00	0.00	0.00	0.00
10	0.00	0.00	0.00	0.00	62.00
15	5.00	13.00	7.00	25.00	77.00
20	4.50	12.50	6.00	23.00	80.00
25	3.00	12.00	6.00	21.00	90.00
30	3.00	11.00	5.00	19.00	85.00
35	0.00	0.00	5.00	0.00	0.00
40	0.00	0.00	0.00	0.00	0.00

C. obtusus

Results are recorded in Table 2. The optimum temperature for *C. obtusus* was 25±1°C. The parasitoid completed the life cycle within 26, 22, 19 and 18.5 days at 15±1°C, 20±1°C, 25±1°C, and 30±1°C, respectively.

Table 2: Effect of Temperature on Development of *C. obtusus.*

Temperature (±1°C)	Incubation Period (Days)	Larval Period (Days)	Pupal Period (Days)	Total Period for Development (Days)	Per cent Survival
5	0.00	0.00	0.00	0.00	0.00
10	0.00	0.00	0.00	0.00	70.00
15	5.00	13.00	8.00	26.00	80.00
20	4.00	11.00	7.00	22.00	85.00
25	3.00	10.00	6.00	19.00	92.00
30	3.00	10.00	5.5	18.50	90.00
35	0.00	0.00	5.00	0.00	0.00
40	0.00	0.00	0.00	0.00	0.00

D. fenestralis

Results are recorded in Table 3. The parasitoid completed its life cycle within 16 days (shortest duration) at 30±1°C. At 5±1°C, 10±1°C, 35±1°C, and 40±1°C, the development was not possible except the pupae at 35°C. The parasitoid completed its life cycle within 22, 20, 18.5 and 16.00 days at 15±1°C, 20±1°C, 25±1°C, and 30±1°C, respectively. The highest survival was observed at 30±1°C. Hence, considered as optimum temperature for development.

Table 3: Effect of Temperature on Development of *D. fenestralis.*

Temperature (±1°C)	Incubation Period (Days)	Larval Period (Days)	Pupal Period (Days)	Total Period for Development (Days)	Per cent Survival
5	0.00	0.00	0.00	0.00	0.00
10	0.00	0.00	0.00	0.00	70.00
15	4.00	11.00	7.00	22.00	78.00
20	3.5	10.00	6.5	20.00	80.00
25	3.00	9.5	6.00	18.50	83.00
30	2.5	8.5	5.00	16.00	91.00
35	0.00	0.00	4.50	0.00	0.00
40	0.00	0.00	0.00	0.00	0.00

Effect of Temperature on the Development and Survival of *Diadegma surendrai* Gupta (Hymenoptera : Ichneumonidae): A Larval Parasitoid of Potato Worm *Phthoromaea operculella* (Zeller) (Lepidoptera)

Abstract

Diadegma surendrai Gupta (Hymenoptera : Ichneumonidae) is an internal larval parasitoid of potato worm Phthoromaea operculella (Zeller) (Lepidoptera : Gelechiidae). The effect of temperature on the development and survival of this parasitoid have been studied at temperatures between 5±1°C to 40±1°C. The development from egg to adult was completed within 26.00, 23.00, 18.00 and 19.00 days at 15±1°C, 20±1°C, 25±1°C and 30±1°C and average percentage of survival was 66.00, 78.00, 80.00 and 60.00 respectively. The optimum temperature for development was 25±1°C.

Keywords: *Temperature, development, Diadegma surendrai.*

Introduction

Temperature affects development, longevity, fecundity, sex-ratio, survival and parasitism by parasitoid (Mellini *et al.,* 1979; Nikam and Sathe, 1981). For mass rearing of parasitoids optimum temperature is crucial factor. Therefore, development of *Diadegma surendrai* Gupta (Hymenoptera : Ichneumonidae) has been studied at different temperature. Review of literature indicates that Beling (1933), Fisher (1959), Nikam and Basarkar (1978 a,b), Nikam and Sathe (1981) etc. attempted the studies related to temperature effect on development of Ichneumonid parasitoids.

Materials and Methods

Four day old larvae of *P. operculella* were exposed to new emerged and mated females of *D. surendrai* in breeding glass cages (size 25 × 25 × 30 cm) for 6 hr for parasitism. Parasitized larvae were transferred into separate containers (size 6.5 × 5.00 cm) for further rearing. Potato tubers were provided to host caterpillars as food during the rearing and 50 per cent honey solution to parasitoid. Effect of temperature on the development and survival of the parasitoid has been studied at 5±1°C, 10±1°C, 15±1°C, 20±1°C, 25±1°C, 30±1°C, 35±1°C and 40±1°C in BOD incubator and seed germinator. Observations on the development of the parasitoid were made at 12 hr interval. Experiments were replicated for five times and in each lot 10 individuals were taken into account.

Results

The results are recorded in Table 4. At 5±1°C, 10±1°C, 35±1°C and 40±1°C the development was not possible. However, at 35±1°C pupal development was possible (6.5 days). The eggs of *D. surendrai* hatched within 5.00, 4.00, 3.00 and 3.00 days; the larval period was 13.00, 12.00, 9.00 and 9.50 days and pupal period was 8.00, 7.00, 6.00 and 6.50 days; thus the life cycle was completed within 26, 23, 18 and 19 days at 15±1°C, 20±1°C, 25±1°C and 30±1°C respectively. The optimum temperature for development of parasitoid was 25±1°C and the average percentage of survival was 66.00, 78.00, 80.00 per cent and 60.00 per cent at 15±1°C, 20±1°C, 25±1°C and 30±1°C respectively.

Table 4: Effect of Temperature on Development of *D. surendrai.*

Temperature (±1°C)	Incubation Period (Days)	Larval Period (Days)	Pupal Period (Days)	Total Period for Development (Days)
5	0.00	0.00	0.00	0.00
10	0.00	0.00	0.00	0.00
15	5.00	13.00	8.00	26.00
20	4.00	12.00	7.00	23.00
25	3.00	9.00	6.00	18.00
30	3.00	9.50	6.50	19.00
35	0.00	0.00	6.50	0.00
40	0.00	0.00	0.00	0.00

Discussion

Beling (1933) studied the development of *Diadegma* sp. at different temperatures and reported that development of *Diadegma* sp. is not possible below 10°C and above 33°C. Similar findings are noted in *Casinario* sp. (Ichneumonidae) by Fisher (1959). While, Nikam and Basarkar (1976 a,b) says that development can be possible even at 35°C. In present form the development was not possible at 5±1°C, 10±1°C, 35±1°C and 40±1°C. However, the parasitoid *D. surendrai* successfully completed its life cycle at 10, 15, 25 and 30°C. At 35°C only pupal development was possible in

D. surendrai. The optimum temperature for larval and pupal development in *C. chlorideae, D. fenestralis* and *D. trichoptilus* were 31°C, 26°C and 25°C respectively (Nikam and Basarkar, 1978 a,b; Nikam and Sathe, 1981) while it was 25±1°C in *D. surendrai.*

According to Nikam and Sathe (1981) adults of *D. trichoptilus* when fed with 20 per cent honey has maximum longevity of 25 days with high percentage of survival at 15°C and the same decreased as temperature increase to 40°C while, *C. chlorideae* and *D. fenestralis* when feed with 20 per cent honey had longevity of 9 days and 24 days at 31°C and 15°C respectively. In *P. testaceous* and *C. obtusus* the survival was maximum at 25±1°C.

Acknowledgements

Author is thankful to UGC, New Delhi for financial assistance to Res. Project F-37-334/2009 (SR)/37-1/2009 (MS) (SR) and Shivaji University for providing facilities.

References

Beling, J. (1933) : Biologie und zucht der schlutwespe *Angitia armillata* Grav. *Arb. boil. Reichsanstalt land Forest Wirtschaft.* Berlin, **20**, 237-244.

Fisher, R.C. (1959) : Life history and ecology of *Horogenes chrysostictes* (Gmelin) (Hymenoptera : Ichneumonidae), a parasitoid of *Ephestia scriacarium* Scott (Lepidoptera : Phycitidae). *Can. J. Zool.*, **37**(1), 429-449.

Mellini, E., Galassi, L. and Briolini, G. (1979) : Effecti della temperature *sulla copia ospite - parasita Galleria mellonella* L. *Gonia cinerascens* Rond. Bullettino dell' Instituto di Entomologia della Universita' di Bologua, 35, 13-28.

Nikam P.K. and Basarkar C.D. (1978a) Studies on the effect of temperature on the development of *Campoletis chlorideae* Uchida (Ichneumonidae), an internal larval parasite of *Heliothis armigera* (Hubn.) *Entomon*, **3**(2), 307-308.

Nikam P.K. and Baraskar C.D. (1978b) : Effect of temperature on *Diadegma fenestralis* (Holmgren) (Hymenoptera : Ichneumonidae), a parasite of *Heliothis armigera* (Hubn.). *Indian J. Parasitol.*, **2**(2), 161-162.

Nikam P.M. and T.V. Sathe (1981) : Studies on the effect of temperature on the development of *Diadegma trichoptilus* Cameron (Hymenoptera : Ichneumonidae), an internal larval parasitoid of *Exelastis atomosa* Walsingham. *Marath. Univ. J. Sci.*, **20**(3), 37-38.

9

Behavioural Aspects

Introduction

Ethology can directly influence the propogation of parasitoids in the laboratory and counts the failure or success of biocontrol programme. Hence, the present topic is devoted to behavioural aspects of Ichneumonid parasitoids namely *P. testaceous, C. obtusus* and *D. fenestralis*. Mating, oviposition and emergence (cocoon construction) behaviour are studied. In past Leius (1961), Dowell and Horn (1975), Gordh and Hendrickson (1976), Lyons (1976), Schimidt (1975), Sathe (1986, 2008), Sathe and Nikam (1983) etc. worked on ethology of Ichneumonid parasitoids.

Materials and Methods

Laboratory culture of parasitoids and their pests have been used for experiments. For mating behaviour newly emerged pair (1 male and 1 female) were caged in test tube size 2.5 × 10 cm and observations were noted every 6 hr per day. At night observations were taken on premating, mating and post mating in red light since parasitoid read Red as Black colour (Sathe and Margaj, 2001). Oviposition was studied by caging individual mated parasitoid female with its host in glass cage (Figures 2 or 3). Similarly, observations on parasitized pests (caterpillars) of respective parasitoids emerging from caterpillars were made at 12 hr interval per day under laboratory conditions (25±1°C, 70-75 per cent R.H. and 12 hr photoperiod) by keeping them in petridishes. The adult emergence from the cocoon also noticed in individual parasitoids. 10 pairs for mating, 20 individuals for oviposition and adult emergence were observed during the course of study.

Results

Results are recorded in Table 1. In all 3 parasitoids, mating behavioural chain comprised as attraction, recognition, orientation, wing farnning, antennation, mounting and copulation. Similarly, in 3 parasitoids oviposition behavioural chain consisted attraction of host, recognition of host, antennal examination of host, up and down movements of abdomen, cleaning ovipositor, thrusting/inserting ovipositor and actual egg laying. All 3 parasitoids, solitary internal, and larval. The last instar (5th) larva in all cases came out from the host body by cutting the body wall of host and then constructed cocoon around itself. Details of behaviours of parasitoids are given in Table 1.

Discussion

According to Dowell and Horn (1975) precopulation period of *B. curculionis* was 35-40 minutes, copulation period was 3.58 minutes and the female was polygamous. Similar observations were noted in *B. anurus* by Gordh and Hendrickson (1976). In the present forms the females were monogamous (mate only once). Sexual receptivity in *B. curculionis* was 48 hours. In the present forms sexual receptivity was 1 to 2 days (avg. 1.5 days). In *Neodiprion sertifer* (Tenthredinidae : Hymenoptera) mating occurred between 9 am to 8 p.m. The present Ichneumonids mated at day time from 7.00 a.m. to 5.00 p.m. According to Sathe (1988) *C. chlorideae* can feed on haemolymph of host occurred due to oviposition trauma. Such type of behaviour was not observed in the present 3 Ichneumonid parasitoids. Oviposition occurred within few seconds in *C. chlorideae* more or less same situation was noted in 3 present parasitoids.

Table 1: Comparative Account on the Mating, Oviposition and Emergence of Adults of Ichneumonids.

Sl.No.	Observations	P. testaceous	C. obtusus	D. fenestralis
(A)	**MATING**			
	i) Precopulation period	1 hr	1 hr	45 minutes
	ii) Copulation period	2 - 3.00 minutes	3-4 minutes	2-3 minutes
	iii) Mating	Male mated 2-3 times in a day Occurred at day, from 7.00 a.m. to 5.00 p.m.	Male had to make several attempts to catch up female, Occurred at day, from 7.00 a.m. to 5 p.m.	Occurred at day, from 7.00 a.m. to 5.00 p.m.
	iv) Feeding after mating	Attracted towards food	Attracted towards food	Attracted towards food
(B)	**OVIPOSITION**			
	i) Pre-oviposition period	1-2 minutes	2 minutes	1-2 minutes
	ii) Oviposition time	2-3 seconds	3-5 seconds	2-3 seconds
	iii) Feeding after oviposition	Attracted towards food.	Attracted towards food.	Attracted towards food.
	iv) Oviposition attempts	Once or twice in a single host	Once or twice in a single host	More than twice in a single host
	v) Superparasitism	No	No	No
	vi) Hyper parasitism	No	No	Yes
(C)	**ADULT EMERGENCE**			
	i) Time required	30-35 minutes	30-40 minutes	20 minutes
	ii) Place of larval emergence	Lateral part of host larva	Lateral part of host larva	Lateral part of host larva.
	iii) Cocoon types and time required for spinning	Solitary, 30-40 minutes	Solitary, 30-45 minutes	Solitary, 20-30 minutes
	iv) Feeding after emergence	Yes	Yes	Yes
	v) Attraction towards opposite sex	Immediately after emergence	Immediately after emergence	Immediately after emergence

For cocoon construction the parasitoid larva reversed several times in all Ichneumonids studied and cocoons were not in uniform colour and size. The cocoons of *C. obtusa* was with penduncle but *D. fenestralis*, *P. testaceous* cocoons were without penduncle. The above aspects will be helpful for finalizing the rearing techniques for parasitoids.

10

Diapause in Ichneumonids

In many insects development is delayed by low humidity in the surrounding air. Eggs of some insects can remain dormant for long periods in a desiccated state. The undeposited eggs of many parasitic Hymenoptera are in a state of arrested growth so long as they stored in the oviducts - since further development is dependent on immersion in the nutrient fluids of the hosts (Flanders, 1942). Voltinism of egg is determined by some influence from the somatic cells of the mother. The type of arrest varies in different species and even among individuals of a single species.

On the basis of diapause the insects are grouped into Heterodynamic and Homodynamic. Diapause is occurred due to adverse conditions as poor food, drought, cold or excessive moisture. The arrest time is vary with species.

After a rest at low temperature, contact with water provides the stimulus for development. Many of the cyclical diapause phenomena of insects are infact induced by seasonal changes. Review of literature indicates that Bodine (1932, 1941), Flanders (1942), Bonnemaison (1945), Bradley and Arbuthnot (1938), Chewyreuv (1913), Marchal (1936), Prebble

(1941), Salt (1941, 1947), Simmonds (1946), Church (1955), Mashennikova (1958) etc. worked on diapause in parasitic insects.

Materials and Methods

Cocoons of *P. testaceous, C. obtusus* and *D. fenestralis* were collected from fields of Brinjal, ber and Jowar from Kolhapur region throughout the year and kept in areated plastic containers in the laboratory of seasonal fluctuating conditions for parasitoid emergence. The fluctuating conditions of the laboratory (temperature, humidity, photoperiod) have been regularly noted for compilation of data generated for diapausing cocoons. Observations were taken on emergence and non emergence of parasitoids from the cocoons and data have been correlated with seasonal conditions and diapause.

Results

Results are recorded in Table 1. *P. testaceous* and *C. obtusus* goes in diapausing stage from January to March every year and *D. fenestralis* from January to August every year. The results recorded in Table 1 indicates that the percent of diapause in *P. testaceous, C. obtusus* and *D. fenestralis* were 5.60, 10.36 and 9.47 respectively. The parasitoid emergence was synchronizing the availability of pest caterpillars in every season.

Table 1: Diapause in Ichneumonids, 2010-2013.

Species	No. of Cocoons Collected	No. of Parasitoids Emerged (Cocoons hatched)	No. of Diapausing Cocoons Emerged in (Next season)	No. of Unhatched Cocoons	Per cent of Diapause
P. testaceous	321	298	18	05	5.60
C. obtusus	357	311	37	09	10.36
D. fenestralis	454	398	43	13	9.47

Discussion

The Hymenopteran phytophagus European spruce sawfly *Gilpinia polytoma* over winters as a larva in the cocoon. Diapause may occur at different stages in the development. Some resting larvae may lie over for seven years in diapausing stage.

Hymenopterous parasitoids specially *Spalangia* and *Cryptus* diapause in the larva is determined by the diet of the mother or by her age at the time of oviposition. None of the larvae from the eggs laid by *Cryptus* female in the first five days of adult life diapaused while, 95 per cent of those from eggs laid 16-20 days after emergence entered into diapause (Simmonds, 1946). The larvae of pteromalid *Mormoniella* sp. enter diapause at the end of the last larval stage if the mother is exposed to low temperatures during oogenesis (Schneiderman and Horowitz, 1958) and as the females become older a greater proportion of the progeny enter diapause at this stage. In some parasitoids slow deposition of the eggs associated probably with partial reabsorption of the yolk may be responsible for causing diapause in the larval stage (Flanders, 1942).

The development of endoparasitic insects is often conditioned by the continued development of the host. If *Plodia* larvae kept under desiccating conditions (29°C, 20 per cent R.H.) the larvae of *Nemeritis* within them remain in the first stage but can develop normally at 26°C and 80 per cent R.H. The Braconid *Cotesia (Apanteles) glomeratus* normally forms clusters of cocoons over the pupating caterpillar *Pieris brassicae* and overwinters as a prepupa but if *P. brassicae* hibernates as a larva, the parasitoid remain inside (Salt, 1941). When *C. (A.) glomeratus* hibernates in diapausing caterpillars of *Aporia crataegi*, appears to be influenced by the physiological changes in the host, but, same species in *P. brassicae* is largely independent on the host. Its diapause was determined by the direct effect of day-length and temperature (Mashennikova, 1958). The parasitoid can therefore, be prevented from entering diapause by exposing to a long photoperiod immediately after emergence from the host. In braconid *Chelonus annulipes* two types of arrest were noticed. If its host *Pyrausta nubilalis* goes into diapause, the parasitoid larva also goes likewise and produced a generation within 2 months in summer but, taken 10 months in winter host.

When *Trichogramma cacoeciae* reared on eggs of diapausing *Cacoecia rosana* its development was inhibited for a period of 7 months but when reared on eggs of *Ephestia cautella* or on eggs of non diapausing *C. rosana* it developed rapidly. It seems that diapause of host can retards the development of parasitoid. In the present study parasitoids have gone into diapause in prepual stage in cocoons and host parasitoid interactions with respect to diapause will be interesting aspect of research as future avenue.

A Dipterous parasitoid *Eucarcelia* was dependent on the great sensitivity of thoracic gland hormone for development in the host. Similarly, a Braconid *Microgaster globatus* dependent on humoral factors in the host. According to Parkar (1935) parasitoid can terminate the diapause earlier than do unparasitized larvae.

The host and parasitoid respond quite independently to the external environment (Doult, 1959). According to Flanders (1942) parasitoids delay the development until the host attains the stage presumably most suitable for the nutritional requirements. While, Simmonds (1947) says that diapause is due to some physiological maladjustment during development and is not an adaptive character. Diapause hormone determines the incidence of diapause. It has been observed that the inhibitory 'diapause factor' are some protein molecules which can be adsorbed upon the activator surface and so cut it off from the pro-enzyme.

However, in parasitic Hymenoptera diapause can occur at every stage of development (Dutt, 1959) and the phenomenon is essential to understand for the survival and field release of parasitoids in biological pest control programmes.

11

Methods of Biological Control

There are four methods of biological control of insects.

1. Collection and destruction of pests by keeping labour.
2. Collection of Ichneumonids from natural habitats and releasing them in target area for pest control.
3. Mass production of Ichneumonids and their release against insect pests in various agro and forest ecosystems.
4. Importation of Ichneumonids, their mass multiplication and use against insects pests.

How to Release Ichneumonids

Ichneumonids should be released in the crop field at evening or morning in adult or cocoon stage. The cocoons are scattered in the field for adult emergence from them and parasitization by Ichneumonids to various kinds of insect pests.

Ichneumonids can be released in two ways for control of insect pests.

1. Small Scale Release

Small number of cocoons or adults are released in the field. Objective of small scale release should be colonization of Ichneumonids and not the immediate control of pests. Long term benefits should be expected from this method.

2. Large Scale Release

Large numbers of cocoons/adults are released in the field for control of pests. Objective of this method may be immediate control of pests and short term benefits.

Ichneumonids Used in Pest Control Programs

1. *Isotima javensis*

It is biocontrol agent of sugar cane top shoot borer *Scirpophaga nivella* and rice stem borer *Triporyza incertulas.* It is released with 115 -120 individuals per hectare thrice from July to August, at weekly interval for good control of sugar cane borers.

2. *Telenomus* sp.

Is used against *Scirpophaga nivella* with 120 individuals per hectare.

3. *Campoletis chlorideae*

Is used against *Helicoverpa armigera* with 100-120 individuals per hectare.

4. *Diadegma argenteopilosus*

It is used against Spodopteran caterpillars with about 100-125 individuals per hectare.

5. *Diadegma semiclausum*

It is used against Spodopteran caterpillars with about 100-125 individuals per hectare.

6. *Xanthopimpla* spp.

These parasitoids are pupal parasitoids and effective against rice borers, jowar stem borer and sugar cane stem borers.

7. *Charop* spp.

Are effective against Ber hairy caterpillars, Pomegranate fruit borers (anar caterpillars) and many other lepidopteran caterpillars.

8. *Goryphus* spp.

Are effective against cotton boll worms.

9. *Diadegma trichoptilus*

It is effective against tur plume caterpillar *Exelastis atomosa.* It may be released 125/ha/week as per the need in tur agro ecosystem.

10. *Goniozus nephantidis*

It is used against coconut pest *Opisina arenosalla* with 20-25 individuals/ha/week along with other parasitoids in coconut ecosystem.

11. *Copidosoma koehlari*

It is effective against potato tuber worm *Phthorimaea operculella* at 1,50,000/ha as per the need.

Contribution to the Society

Ichneumonids are biopesticides scattered in environment. They should be collected, identified and evaluated and used in eco-friendly pest control measures specially biological pest control. 40 Ichneumonids have been identified and 3 Ichneumonids have been evaluated with respect to their reproductive potential. Therefore, the present study will be very useful for designing eco-friendly pest control and keeping safeguard environment, flora, fauna and human community as a whole.

12

Research Papers

(1) Biology of *Diadegma trochanterata* (Morley) (Hymenoptera: Ichneumonidae): An Internal Larval Parasitoid of the Caster Capsule Borer *Dichocrocis punctiferalis* Guene (Lepidoptera : Pyralididae)

T.V. Sathe

Department of Zoology, Shivaji University, Kolhapur – 416 004, India

ABSTRACT

Diadegma trochanterata (Morley) (Hymenoptera : Ichneumonidae), is an internal larval parasitoid and biocontrol agent of the caster capsule borer Dichocrocis punctiferalis Guene (Lepidoptera : Pyralididae). Hence, objecting its utility in biological control of D. punctiferalis biology of D. trochanterata has been studied. The parasitoid completed its life cycle from egg to adult within 18.5 days. Adult longevity averaged 15 days and 17 days in males and females respectively with 20 per cent honey. Mating occurred at day time, oviposition takes place within 40-50 seconds. The parasitoid selected 4-5 day old larvae of the host for maximum parasitism, 46.00 per cent under laboratory conditions (25 ±1°C, 70-75 per cent R.H. and 12 hr photoperiod).

Keywords: *Diadegma trochanterata, Parasitoid, Biology, Dichocrocis punctiferalis, Caster pest.*

Introduction

Diadegma trochanterata (Morley) (Hymenoptera : Ichneumonidae) is an internal, larval parasitoid of the caster capsule borer *Dicrococis punctiferalis* Guene (Lepidoptera : Pyralididae). Caster *Ricinus communis* L. is oil seed and sericultural crop. The leaves of this crop are used to feed eri silkworms *Samia cynthia ricini*. Hence, control of this pest is essential part from the view of agriculture and sericulture. Chemical control is not without danger to silkworm and secondly, pesticides lead several serious problems like pollution, pest resistance, pest resurgence, secondary pest out break etc. Keeping in view all above facts, present work was carried out. Review of literature indicates that Atwal (1976), Fisher (1959), Gangrade (1964), Tikar and Thakre (1961), Leong and Oatman (1968), Bartell and Pass (1978), Sathe (1988, 1990, 2008) etc. attempted biological studies in Ichneumonid parasitoids of various insect pests.

Materials and Methods

The larvae of *D. punctiferalis* and cocoons of *D. trochamtera* have been collected from the fields of caster to maintain laboratory culture of parasitoid. Similarly, host larvae also reared in the laboratory on caster capsule. 4-5 day old 50 larvae of *D. punctiferalis* were exposed to two mated females of parasitoid in glass cage (25 × 25 × 30 cm) for 6 hr. The parasitized larvae were dissected after every 12 hr in normal saline solution and parasitoid eggs and larvae were collected for further observations. The larval stages were treated with 50 per cent chloroform and 50 per cent ethanol and mounted in Hoyer's medium on microslides for biometrical studies. Newly emerged sexes were kept separate in specimen tubes and confined in a pair (1 male and 1 female) into the test tube (10 × 2.5 cm) for mating studies. Similarly, a host larva was taken into test tube and then mated female was released for noting the observations on oviposition. Host age selection was studied by exposing 50 host larvae ranging from 1 day old to 20 day old in glass cage (25 × 25 × 30 cm) for

12 hr and the parasitoid larvae were then separated and reared on caster capsules under laboratory (25±1°C, 70-75 R.H.; 12 hr photoperiod). Per cent of parasitism was noted on each day on the host larvae.

Results

Life Cycle

The parasitoid showed five larval instars. First two were caudate and remaining three were hymenopteriform. The life cycle period from egg to adult formation was 18.5 days. The parasitoid showed 4 distinct stages of life cycle *viz.* egg, larva, pupa and adult.

Egg

Eggs were elongately curved and tapered at one end and rounded at another end. The chorion was smoothly upaque and white. Incubation period was 3.5 days. Mostly single egg was laid by the parasitoid on host body but rarely 2 to 3 eggs also laid. In 20 individuals eggs averaged 0.26 mm and 0.062 mm in length and width.

First Instar

First instar larva was creamy white and with long tail *i.e.* caudate type. The tail length was about the half the length of the body. 13 body segments were visible but not clear distinction between thorax and abdomen. The tracheal system was with single longitudinal trunk with several branches. Spiracles were not prominent. The parasitoid consumed internal tissues of the host larva. This stage lasted for 3 days and averaged 1.25 mm in length, 0.16 mm in width and mandibles averaged 0.20 mm in length and 0.014 mm in width.

Second Instar

This stage was opaque with reduced tail and prominent 13 well defined body segments. No vesicle was present. The tracheal system showed two longitudinal trunks with small branches. Spiracles were poorly seen. The larva averaged 1.95 mm in length and 0.36 mm in width. Mandibles in 20 individuals averaged 0.28 mm in length and 0.020 mm in width. The larva was typically curved. This stage lasted for 2 days during which the larva consumed internal tissues of the host larva keeping intact important organs of the host.

Third Instar

Third instar larvae were yellowish opaque in body colour and measured 3.00 mm in length 0.42 mm in width in 20 individuals. Mandibles averaged 0.038 mm in length and 0.027 mm in width. This stage lasted for one day. The tracheal system was with two longitudinal trunks with well developed several branches of tracheae. Spiracles were prominent.

Fourth Instar

This stage was with yellowish opaque coloured and with well defined body segments. Fourth instar was longer than fifth instar and more straight than other instars. No trace of tail was present. The tracheal system was with transverse trunk and two longitudinal trunks with well developed side branches. Spiracles were very prominent. The larvae averaged 4.00 mm in length and 0.73 mm in width and mandibles 0.068 mm in length 0.034 mm width in 20 individuals. This stage lasted for 2 days and larva consumed host tissues.

Fifth Instar

Fifth instar was also yellowish opaque in colour but was curved and slightly tapered to both the ends and was typically hymenopteriform. The larva consumed internal tissues of the host body and full grown in about 2.5 days. Tracheal system was more developed with more prominent spiracles. The larvae averaged 3.6 mm in length and 0.76 mm in width and mandibles averaged 0.070 mm in length 0.055 mm in width in 20 individuals.

Cocoon

Cocoon was dirty white coloured, rounded at both the ends and averaged 4.2 mm in length and 1.10 mm in width. The parasitoid emerged from anterior side of the cocoon by taking circular cut to the cocoons.

Pupa

Pupa was exarate type, whitish in body colour at initial stage but became dark brown approaching its full development. Pupae averaged 3.8 mm and 0.76 mm in length and width respectively in 20 individuals. Pupal duration averaged 6 days.

Adult

Adult males were smaller than females and females with ovipositor

96 *Biological Pest Control through Ichneumonids*

for eggs laying. The ovipositor was curved upward and was not straight. Head, thorax and some part of dorsal side of abdomen was black in colour.

Adult Longevity

The males and females survived for an average of 15 days and 17 days respectively with 20 per cent honey. In control, the parasitoid died within 2 days and with water the parasitoid extended survival for 3 days.

Mating

Mating occurred at day time, about afternoon. Mating period averaged 2.45 min.

Oviposition

Parasitoid oviposited 4.5 days old host larvae immediately when confined in test tubes. Oviposition was completed with 35 seconds.

Host Age Selection

The parasitoid selected 4-5 day old host larvae for maximum progeny production (46.00 per cent). The hosts 1-2 days old and 15 to 20 day old were rejected by the parasitoid for parasitism. The progeny production was possible from hosts of 3 day old to 12 day old hosts. The sex ratio was favouring females.

Discussion

Fisher (1959) reported 4 instars in *Horogenes chrysostictes* (Gemelin) (Ichneumonidae), a parasitoid of *Ephestia sericarium* (Scott.). Similarly, Tikar and Takre (1961) reported 4 instars in *H. fenestralis* Holmgren, a parasitoid of common caterpillar. While, Gangrade (1964) and Sathe (2008) reported five instars in *Campoletis oblorideae* Uchida, an internal larval parasitoid of *Helicoverpa* (=*Heliothis) armigera* (Hubn.). In an Ichneumonid parasitoid *Diadegma argenteopilosa* (Cameron), a parasitoid of *Spodoptera litura* (Fab.) also reported 5 instars. The present findings are in agreement with number of instars reported by Gangrade (1964) and Sathe (2008) in respective parasitoid species. The life cycle from egg to adult was completed within 17 and 18 days in *D. trichoptilus* (Cameron) and *D. argenteopilosa* (Cameron) respectively (Sathe 1988; Sathe, 2008) while in the present form *D. trochanterata* the life cycle was completed with 20 days. *D. trochanterata* mate and oviposit immediately in the laboratory and selects 4-5 day old caterpillars for maximum parasitization. The above attributes of *D. trochantera* will add great relevance in designing mass rearing technique for the same parasitoid in future.

Acknowledgement

Author is thankful U.G.C., New Delhi for providing financial assistance to the UGC Major Project F. No. 37-334/2009 (SR).

References

Atwal, A.S. 1976. Agricultural pests of India and South East Asia. Kalyani Publ., New Delhi. p. 303.

Bartell, D.P. and Pass, B.C. 1978. Morphology, development and behaviour of the immature stages of the parasite *Bathyplectes curculionis. Ann. ent. Soc. Am.;* 7(1), 23-30.

Fisher, R.C. 1959. Life history and ecology of *Horogenes chrysostictes* (Gemelin) (Hymenoptera : Ichneumonidae), a parasitoid of *Ephestia sericarium* (Scott.) (Lepidoptera : phycitidae). *Can. J. Zool., 37,* 429-446.

Gangrade, G.A. 1964. On the biology of *Campoletis perdistinctus* Uchida (Hym; Ichneumonidae) in Madhya Pradesh. *Ann. ent. Sco. Am;* 231-244.

Leong G.K.L. and Oatman E.R. 1968. The biology of *Camplex haywardi* (Hym. :Ichneumonidae), a primary parasitoid of the potato tuber worm. *Ann. ent. Soc. Am.* 6, 26-36.

Sathe, T.V. 1988. Biology of *Diadegma trichoptilus* (Cameron) (Hymenoptera: Ichneumonidae), a larval parasitoid of *Exelastis atomosa* Walsingham. *J. Curr. Biosci;* 5(2), 37-40.

Sathe, T.V. 1990. The Biology of *Diadegma trichoptilus* (Cameron) (Hymenoptera : Ichneumonidae), an internal larval parasitoid of *Spodoptera litura* (Fab.) *The Entomologist,* 109(1), 2-7.

Sathe, T.V. 2008. Mass production technique for *Compoletis chloridae* Uchida. *Biotechnological Approaches in Entomology;* 3, 64-74.

Sathe, T.V. and A.D.Jadhav 2001. Sericulture and Pest Management Daya publ. House, New Delhi. pp. 1-197.

Tikar, D. T. and Thakre, K.R. 1961. Bionomics and biology of immature stages of an Ichneumonid *Horogenes fenestralis* Holmgren, a parasitoid of common caterpillar. *Indian J. Ent.;* 23, 116-124.

(2) The Tracheal System in Immature Stages of *Diadegma trichoptilus* (Cameron) (Hymenoptera : Ichneumonidae): A Parasitoid of Tur Plume Caterpillar *Exelastis atomosa* Walsingham (Lepidoptera : Pterophoridae)

T.V. Sathe

Department of Zoology, Shivaji University, Kolhapur – 416 004, India

ABSTRACT

Diadegma trichoptilus (Cameron) (Hymenoptera : Ichneumonidae) is an internal larval parasitoid of Tur plume caterpillar Exelastis atomosa Walsingham (Lepidoptera : Pterophoridae) and acts as good biocontrol agent. Therefore, as a base line data, trachal system of immature stages of this parasitoid has been studied. There are five instars in the parasitoid. In first instar trachal system comprised a thin longitudinal trunk running from anterior to posterior end with lateral branches. In second instar lateral branches were more thicker than first instar. In 3ʳᵈ, 4ᵗʰ and 5ᵗʰ instar larvae the trachal system was much more branched and complicated. 6 pairs of spiracles were locked in first 6 segments of the body. A thicker network of tracheae were found on the body. This work will be helpful for identification of instars in an endoparasitoid.

Keywords: *Diadegma trichoptilus, Parasitoid, Tracheal system, Identity of insect, Exelastis atomosa.*

Introduction

Diadegma trichoptilus (Cameron) (Hymenoptera : Ichneumonidae) is an internal larval parasitoid of plume caterpillar *Exelastis atomosa* Walsingham (Lepidoptera : Pterophoridae). It parasitizes second instar caterpillar and acts as good biocontrol agent. Being an internal larval parasitoid counting and identifying instars in host larva is difficult phenonmenon. The study of tracheal system has great importance from the view point of taxonomy. Few reports are available on respiratory system of parasitoids (Seurat, (1898), Thorpe, (1930), Farooqi *et al.,* 1965, Ingawale and Sathe, 1994, Sathe and Dawale, 2001 etc.) However, very little attention is paid on Ichneumonid parasitoids. Keeping in view above facts present work has been carried out.

Materials and Methods

Tur plume caterpillars *(E. atomosa)* were collected from the fields of Kolhapur region and reared in the laboratory (25±1°C, 75-80 per cent R.H. and 12 hr photoperiod) by providing its natural food (Red gram) for screening parasitoids. For mating, newly emerged pair (♂ and ♀) was confined into plastic container of 1 lit. size perforated with small exits for aeration to insects. Mated female was allowed to lay eggs on second instar *E. atomosa* larva. Later, parasitized larvae were dissected for collection of parasitoid larvae. Sufficient number of larvae of every instar were collected (Sort, 1959) and respiratory system was observed under compound microscope and measurements of essential parts were taken in mm.

Results

D. trichoptilus showed 5 larval instars. Out of which first two were caudate type and rest hymenopteriform. In the first instar tracheal system showed a thin longitudinal trunk running from anterior to posterior end with lateral branches. Two pairs of spiracles were located between 1st and 2nd body segment at the tip of lateral trunk. The spiracles were nonfunctional and tracheae were filled with fluid. In the 2nd instar lateral branches were more thicker than first instar. The tracheae were filled with gas and 4 pairs of spiracles were present. In 3rd, 4th, and 5th instar larvae the tracheal system was much more branched and filled with gas. Above instars showed 8 pairs of spiracles, first two pairs situated in meso and meta thorax and 6 spiracle pairs were in first 6 abdominal segments, one pair in each segment. Last three instars showed two longitudinal tracheal trunks and dorsally a anterior and a posterior commissures. Approaching final and matured instar, lateral tracheae developed more complex structure.

Discussion

The respiration of endoparasitic insect larvae shows some striking parallels with the respiration of aquatic forms (Wigglesworth, 1972). The tracheae in first instars were nonfunctional in the agromyrid parasitoid *Cryptochaetum* sp. Therefore, the exchange of gases takes place directly between the tissue fluids of parasitoid and host and the spiracles were non functional. Similar results are noted in the present parasitoid. According to Seurat (1898) from 2nd instar onwards spiracles became functional in *Nemeritis* sp. a parasitoid of *Ephestia cautella* caterpillar. When tracheae fill with gas, in 2nd instar, they supply a rich network of fine branches to the skin and facilitated the larva upto emergence from its host.

Farooqui *et al.* (1965) studied a trachal system in *Devorgilla inquinata* Morley (Ichneumonidae) wherein they noted two pairs of spiracles in first instar, 4 pairs in second instar and 7 pairs from 3rd to 5th instars. Farooqui *et al.* (1965) used the size of the tail and the extent of development of the tracheal system for identification of instars in the parasitoid.

In an Ichneumonid, *Diadegma argenteopilosa* Cameron, a parasitoid of *Spodoptera litura* (Fab.) two longitudinal tracheal trunks have been reported by Sathe and Dawale (2001). They further noted anteriorily two trunks connected just below the head capsule by a dorsal commissure and a posterior commissure was also visible but upto 3rd instar spiracles were not clear. However, in the fifth instar they noted 9 pairs of spiracles, 7 pairs in abdominal region and 2 pairs in thoracic region. In the present parasitoid 6 pairs of spiracles were present on abdominal first six segments and 2 were on thoracic segments. Ingawale and Sathe (1994) studied tracheal system in a braconid *Apanteles jayanagarensis* Bhatnagar, a larval parasitoid of *Spilosoma obliqua* Walker and noted a single longitudinal tracheal trunk in first instar. Similarly, second instar also showed single longitudinal trunk but, with well developed lateral tracheal branches. However, in third instar the tracheal system was with 2 longitudinal trunks and 6 pairs of spiracles, 2 pairs on thorax and 6 pairs on first 6 abdominal segments.

In the present parasitoid first two instars were caudate type and remaining 3 were hymenopteriform. In first instar, only single branch of trachea was extended in the tail while in second instar well developed lateral branches of tracheae were recorded in the tail. In Hymenopteri form, two longitudinal tracheal trunks were observed. Very recently, Dubey and Paliwal (2006) studied the tracheal system of sheep body louse *Damalinia ovis* (Linn.) where in they reported two main tracheal trunks, one on each side of the body and 8 pairs of spiracles six pairs in abdominal segments and 2 pairs in thoracic segments. In the present form same situation was noticed.

Acknowledgement

Author is thankful to U.G.C., New Delhi for providing financial assistance to Major Research Project F. No. 37-334/2009 (SR) and Shivaji University, Kolhapur for providing facilities.

References

Dubey R.K. and A.K. Paliwal 2006. The tracheal system of sheep body louse *Damalinia ovis* (Linn.) (Mallophaga). *Uttarpradesh J. Zool.*, **26**(2), 239-241.

Farooqui S.I., B.R. Subba Rao and A.K. Sharma 1965. Studies on the parasites of *Orthaga* sp.; a pest of *Syzygium fruticosum* Roxb. at Delhi. *Beitr. Zur. Entom. Band.;* **15**(1/2), 179-196.

Ingawale D.M. and T.V. Sathe 1994. Biology and biometry of immature stages of *Apanteles jayanagarensis* Bhatnagar (Hymenoptera : Braconidae), an endoparasitoid of *Spilosoma obliqua* (Walker) *J. Anim. Morph. and Physiol.* **41,** 13-17.

Sathe, T.V. and R.K. Dawale 2001. Morphology and biometry of *Diadegma argenteopilosa* Cameron (Hymenoptera : Ichneumonidae), a parasitoid of *Spodoptera litura* (Fab.) (Lepidoptera). *J. Ent. Res.*, **25**(1), 103-107.

Seurat L.G. Respiration of parasitic larvae, Hymenoptera. *Ann. Sci. Nat. Zool.*, **10**, 1-159.

Short, J.R.T. 1959. A description and classification of final instar larvae of Ichneumonidae (Insecta : Hymenoptera). *Proc. U.S. Nat. Mus.* 110, 391-511.

Thorpe W.H. 1930. Respiration of parasitic larvae, *Cryptochaetum iceryae. Proc. Zool. Soc. Lond*, 927-71.

Wigglesworth, V.B. 1972. The principles of insect physiology. Pp. 385-388.

(3) Reproductive Potential of *Diadegma insulare* (Cameron) (Hymenoptera : Ichneumonidae) in Relation to Age of Diamond Back Moth *Plutella xylostella* (Linnaeus) (Lepidoptera : Plutellidae)

T.V. Sathe

Department of Zoology, Shivaji University, Kolhapur – 416 004, India

ABSTRACT

Diadegma insulare (Cameron) (Hymenoptera : Ichneumonidae) is an internal larval parasitoid of Diamond back moth Plutella xylostella (Linnaeus) (Lepidoptera : Plutellidae), a serious pest of cabbage and canliflower in India. Age of the host plays an important role in successful parasitism and causing mortalities in pest species. Hence, the present study was objected to find out most suitable age of P. xylostella for maximum progeny production. The experiments conducted with age 1 to 16 days hosts at laboratory conditions (27 ± 1°C, 80 R.H. and 12 hr photoperiod) showed that 4 day old caterpillars were most suitable for maximum progeny production. 5 to 10 day old hosts were accepted by the parasitoid but comparatively yielded less progeny.

***Keywords**: Diadegma insulare, Host age selection, Plutella xylostella.*

Introduction

The host parasitoid association is a complex phenomenon (Vinson and Iwantsch, 1980). In mass production, release and colonization of parasitoid host shape, size, movement, sound, nutrition, immunosuitability and age plays very crushial role individually or in combinations (Putter and Vandan Bosch. 1959; Richerson and Deloach 1972; Price, 1970). Review of literature indicates that Leong and Oatman (1968), Lingren *et al.* (1970), Sathe and Nikam (1985), Sathe (1990), Sathe and Jadhav (2008) etc. attempted host age selection in Ichneumonid parasitoids.

Materials and Methods

50 larvae of *P. xylostella* of known age, ranging from 1 day to 16 day old were exposed to a single mated female of *D. insulare* in glass case of size 25 cm × 25 cm × 25 cm (L × W × H) for 24 hr for parasitization. The parasitized larvae then removed to separate containers for further developmental studies. Daily records of parasitoid emergence/moth

emergence from each lot were made and per cent parasitization per 100 larvae was calculated. During the experiments the pest was fed with cabbage bulbs and parasitoid with 50 per cent honey solution. Each experiment was replicated for five times for confirming the results.

Results

Results recorded in Table 1 indicate that the parasitoid produced highest progeny production with 4 day old host larvae. The young larvae 1 day old and beyond 15 day old were not parasitized by the parasitoid. However, the larvae and 3 to 10 day old were readily accepted by parasitoid for parasitization. The observations were also made on sex ratio of progeny produced. The sex ratio was mostly in favour of females in older larvae and in males in younger larvae.

Table 1: Host Age Selection by *D. insulare.*

Host Age in Days	Total No. of Hosts Tried	Total No. of Parasitoids Emerged			Per cent Parasition
		Male	*Female*	*Total*	
1	250	0.0	0.0	0.0	—
2	250	26	24	52	20.00
3	250	53	51	104	41.30
4	250	54	56	110	44.00
5	250	52	56	108	43.20
6	250	50	50	100	40.00
7	250	40	56	96	38.40
8	250	40	42	82	32.80
9	250	39	41	80	32.00
10	250	40	40	80	32.00
11	250	32	33	65	26.00
12	250	25	27	52	20.80
13	250	24	24	48	18.00
14	250	14	16	30	12.00
15	250	3	5	8	3.20
16	250	0	0	0	0.00

Discussion

Lingren *et al.* (1970) studied the pest age preference given by *Campoletis chlorideae* Uchida towards Army worm *Pseudoletia unipuncta* (Haworth),

Cabbage looper *Trichoplusiani* (Hubner), the southern army worm *Prodenia eridinia* (Cramer) and the western yellow stripped army worm *Pseudoletia praefica* Grote. Their observations indicated that 1-8 day old pest larvae were susceptible for parasitism but in general 2 to 6 day old larvae were readily parasitized and 2 to 4 day old larvae were most acceptable.

Leong and Oatman (1968) studied the relationship between host age and the rate of parasitization in *Campoplex haywardi* Blanchard (Ichneumonidae), a parasitoid of potato tuber worm *Pthorimaea operculella* Zeller. They provided 50 potential host larvae/tuber of various host ages for parasitization and observed that 5 to 6 day old tuber moth larvae yielded the highest average number (13.9) of parasitoids. In the present study also host larvae were exposed to parasitoids with 50 host density but, parasitoid preferred 4 day old host larvae for maximum progeny production. Food for both, the host and parasitoid was provided in sufficient quantities.

According to Oatman and Platner (1974) *Temelucha* sp. *platensis* group (Hymenoptera : Ichneumonidae) preferred 3 - 4 day old caterpillars of *P. operculella* while, Sathe and Nikam (1985) recorded highest per cent parasitism (21.33) made by *Diadegma trichoptilus* (Cameron) in caterpillars of *Exelastis atomosa* Walsingham (Lepidoptera : Pterophoridae). However, beyond 9 day old host, no parasitization was recorded. More or less same pattern of preference was given by *D. insulare* towards *P. xylostella.*

Sathe and Jadhav (2008) studied host age selection by *Diadegma ricini* Row and Kurian in *Dichocrocis punctiferalis* Ewenee. Wherein they noted maximum 48 per cent parasitism on 3-4 day old host larvae. 1 day and 8 to 10 day old host larvae were not selected by the parasitoid for parasitism.

Effect of age of the host on parasitization in *Pseudapanteles dignus* (Muesebeck) (Hymenoptera : Braconidae), a parasitoid of *Keiferia cycopersicella* (Walsingham) at insectary conditions was observed by Cardona and Oatman (1971) wherein 2 to 3 day old host larvae were most suitable for parasitization, since then the highest per cent of parasitization (48.1 per cent) and the highest number of parasitoids (12) were obtained with this age. Very interestingly, the progeny production decreased gradually with an increase in the age of the host larvae. In the present study also same trend of progeny production was found beyond

4 day old host larvae. However, in older host instars sex ratio was in favour of females in the present study noting the probable need of nutritional requirement. Investigations on pest stage attack by parasitoid are wanting for fruitful exploitation of parasitoids in biological pest control.

Acknowledgements

The author is thankful to UGC, New Delhi for sanction of the project, - No. 37/334, 2009 (SR) for financial assistance and Shivaji University for providing facilities.

References

Cardona, C. and Oatman E.R. 1971. Biology of *Apanteles dignus* (Hymenoptera : Braconidae), a primary parasite of Tomato pin worm. *Ann. ent. Soc. Am.*, 64, 996-1007.

Leong G.K.L. and Oatman E.R. 1968. The biology of *Campoplex haywardi* (Hymenoptera : Ichneumonidae), a primary parasite of the potato tuber worm. *Ann. ent. Soc. Am.*, 61, 26-36.

Lingren P.B., Guerra, R.J., Nickelson J.W. and White, C. 1970. Hosts and host age preference of *Campoletis perdstinctus. J. Econ. Ent.,* 63, 518-522.

Oatman E.R. and Platner G. R., 1974. The biology of *Temelucha* sp. *platensis* group (Hym. : Ichneumonidae), a primary parasite of the potato tuberworm. *Ann. ent. Soc. Am.*; 67, 276-280.

Price P.W. 1970. Trial odors : recognition by insect parasitic on cocoons. *Science,* 170, 546-547.

Puttler, B. and Vanden Bosch, R. 1959. Partial immunity of *Laphygma exigua* (Hubner) to the parasite *Hyposoter exigue* (Viereck). *J. Econ. Ent.*, 52, 327-329.

Richerson J.V. and Deloach C.J. 1972. Some aspects of host selection of *Perilitus coccinellae. Ann. ent. Soc. Am.*, 65, 834-839.

Sathe T. V. and B. V. Jadhav, 2008. Host age selection by *Diadegma ricini* Row and Kurian (Hymenoptera : Ichneumonidae) in caster capsule borer *Dichorocis punctiferalis* Gwen. (Lepidoptera : pyralididae). *Biotech. Approach. Ent.*, 3, 75-80.

Sathe, T.V. 1985. Studies on the host age selection by *Diadegma trichoptilus* (Cameron), a larval parasitoid of *Exelastis atomosa*) Wals. *Curr. Sci.* 54(15), 762—68.

Sathe, T.V. 1990. The biology of *Diadegma argenteopilosa* Cameron (Hym. : Ichneumonidae), an internal larval parasitoid of *Spodoptera litura* (Fab.). *The Entomologist*, 109, 2-7.

Vinson S.B. and Iwantsch G.B. 1980. Host suitability for insect parasitoids. *A. Rev. Ent.*, 25, 397-419.

(4) Reproductive Potential of *Goryphus nursei* Cameron (Hymenoptera : Ichneumonidae) with Respect to Age of its Host *Sylepta derogata* Fabricius (Lepidoptera : Pyraustidae)

T.V. Sathe and Nilam Shendge

Department of Zoology, Shivaji University, Kolhapur – 416 004, India

ABSTRACT

Goryphus nursei Cameron (Hymenoptera : Ichneumonidae) is an internal larval parasitoid of the cotton leaf roller Sylepta derogata Fabricius (Lepidoptera : Pyraustidae) and acts as good biocontrol agent. Therefore, reproductive potential of G. nursei has been studied with respect to host age. The parasitoid selected 5 day old S. derogata larvae for maximum progeny production. The larvae 1 to 3 day old and 12-20 day old were not accepted for parasitism. Other host ages produced less progeny production.

Key words: *Goryphus nursei, Parasitoid, Progeny, Sylepta derogata, Cotton leaf roller.*

Introduction

Goryphus nursei Cameron (Hymenoptera : Ichneumonidae) is an internal larval parasitoid of the cotton leaf roller *Sylepta derogata* Fabricius (Lepidoptera : Pyraustidae) and acts as a good biocontrol agent. Various aspects of hosts are responsible for successful parasitism (Vinson and Iwantsch, 1980). High rate of percentage of parasitism is most desirable feature of an ideal parasitoid (Sathe and Margaj, 2001). The host parasitoid association is complex phenomenon and host age, shape, size, movement, sound and nutrition play a very crucial role in host selection and successful parasitism. The detection of pest age attacked by natural enemy is an important aspect for release of biocontrol agents in the field at appropriate time and stage of the pest insect. The review of literature indicates that Lingren *et al.* (1970), Leong and Oatman (1968), Oatman and Platner (1974), Sathe (1990), Sathe and Nikam (1985), Vinson and Iwantsch (1980), Sathe and Margaj (2001) etc. attempted the work related to host age selection by Ichneumonid parasitoids.

Materials and Methods

The cultures of parasitoid and its host were maintained in the laboratory by collecting cocoons of parasitoids and larvae of hosts from fields. To determine the effect of host age on progeny production, 50 host larvae of known age (range 1 day to 20 day old) were exposed to single mated female of *G. nursei* in an oviposition unit for 12 hr. Following exposure, the larvae were removed to separate containers for further observations. Daily records of parasitoid emergence from each lot were made. Each experiment was replicated 5 times. 20 per cent honey and cotton leaves were given to parasitoid and host during the conduct of experiments.

Results

Results are recorded in Table 1. The highest number (127) of parasitoids were emerged from host age group 5 day old. The host larvae 1 to 2 day old and 12 to 20 day old were remained unparasitized. However, the host age 5 to 12 and 3-4 day old were also yielded progeny production. In general, 4 to 10 day old host larvae have yielded high number of parasitoids.

Discussion

Lingren *et al.* (1970) studied the host age preference of *Campoletis chlorideae* Uchida *(= Campoletis perdistinctus)* towards the four lepidopterous host species *viz.* army worm *Pseudoletia unipuncta* (Haworth), Cabbage looper *Trichoplusia ni* (Hubner), the southern armyworm, *Prodenia eridinia* (Cramer) and the western yellow stripped armyworm, *Pseudoletia praefica* Grote. Their results indicated that 1-8 day old host larvae were succeptible for parasitism but, in general, 2-6 day old were readily parasitized and 2-4 day old larvae being most acceptable.

Leong and Oatman (1968) provided 50 potential host larvae/tuber of various host age for parasitization and observed that 5 to 6 day old tuber

Table 1: Effective Age of *S. derogata* Larvae for Progeny Production by *G. nursei.*

Sl.No.	Age Days	No. of Moth Emerged	Total Number of Parasitoids Emerged			Parasitism Per cent	Mortality Per cent
			Male	Female	Total		
1	1	230	—	—	—	—	10.50
2	2	235	—	—	—	—	9.70
3	3	225	6	9	15	6.00	9.00
4	4	115	51	69	120	48.00	6.00
5	5	110	54	73	127	50.80	5.20
6	6	112	55	67	122	48.80	6.4
7	7	124	50	60	110	44.00	6.4
8	8	136	42	50	92	36.80	8.00
9	9	150	38	42	80	32.00	8.00
10	10	180	20	38	58	23.20	4.80
11	11	192	15	21	36	10.44	8.00
12	12	230	—	—	—	—	10.5
13	13	234	—	—	—	—	9.36
14	14	232	—	—	—	—	9.28
15	15	236	—	—	—	—	9.44
16	16	232	—	—	—	—	9.28
17	17	232	—	—	—	—	9.28
18	18	235	—	—	—	—	9.7
19	19	235	—	—	—	—	9.7
20	20	236	—	—	—	—	9.44

moth larvae *Phthorimaea operculella* Zeller yielded the highest number of progeny. While Oatman and Platner (1974) noted 3-4 day old larvae of *P. operculella* as the most suitable for parasitization by *Temelucha* sp. of *platensis* group but the per cent parasitism was slightly higher with 2 to 3 day old host larvae. In the present study, *G. nersei* showed maximum progeny production on 5 day old host larvae. The larvae 1 to 2 day old and 12 to 20 day were not suitable for progeny production. However, the host age group 3 and 11 day old were susceptible for parasitism but yielded comparative less number of parasitoids. Sathe and Nikam (1985) recorded optimum 2-3 day old age of *Exelastis atomosa* (Wals.) for maximum progeny production by *Diadema trichoptilus* while, in present study *G. nursei* showed greatest response to 5 day old host larvae for maximum progeny production. The present data will be helpful for mass rearing of *G. nursei* and further for biological control of *S. derogata*.

Acknowledgement

Author is thankful to U.G.C., New Delhi for providing financial assistance to the UGC Major Project F. No. 37-334/2009 (SR).

References

Leong G.K.L. and Oatman E.R. 1968. The biology of *Complex haywardi* (Hymenoptera : Ichneumonidae), a primary parasite of the potato tuber worm. *Ann. ent. Soc. Am.*; 61(1), 26-36.

Lingren P.D., Guerra R.J., Nickelson J.W. and C. White, 1970. Hosts and host age preference of *Campoletis perdistinctus. J. Econ. Ent.*; 63, 518-522.

Oatman E.R. and Platner G.R. 1974. The biology of *Temelucha* sp. *platensis* group (Hymenoptera : Ichneumonidae), a primary parasite of the potato tuber worm. *Ann. ent. Soc. Am.*; 67, 275-280.

Sathe T.V. 1990. The biology of *Diadegma argenteopilosa* Cameron (Hymenoptera : Ichneumonidae), an internal larval parasitoid of *Spodoptera litura* (Fab.). *The Entomologist* (U.K.), 109, 2-7.

Sathe T.V. and P.K.Nikam 1985. Studies on the host age selection by *Diadegma trichoptilus* (Cameron), a larval parasitoid of *Exelastis atomosa* (Wals.). *Curr. Sci.*, 54(15), 262-263.

Sathe T.V. and G.S. Margaj 2001. Cotton pests and biocontrol agents. Daya Publ. House, Delhi. pp.1-166.

Vinson S.B. and G.B. Iwantsch 1980. Host suitability for insect parasitoids. *A. Rew. Ent.*, 25, 397-419.

Bibliography

Basarkar C.D. and P.K.Nikam 1981. Longevity, fecundity and Sex-ratio of *Goryphus nursei* (Cameron) (Hymenoptera : Ichneumonidae), a solitary parasitoid of *Earias vitella* Stoll (Lep., Arctiidae), *Z. ang Ent.*, 89-91.

Bodine J.H. 1932. Mechanism of diapause. *Physiol. Zool.* 5, 549-554.

Bonnemaison, L. 1945. Diapause: Review. *Ann. Epiphyt.* 11, 19-56.

Bradley W.G. and K.D. Arbuthnot 1938. Effect of host (*Pyrausta*) on diapause in *Chelonus*, Braconidae. *Ann. ent. Soc. Am.*, 31, 359-365.

Cameron, P. 1891. On some Hymenoptera parasitic on Indian injurious insects. *Mem.Proc. Manchr. Lit. Phil. Soc.* 4(4), 1-185.

Cameron, P. 1905. On the phytophagus and parasitic Hymenoptera collected by Mr. E. Ernest Green in Ceylon. *Spolia Zeylan*, 3, 67-143.

Chewyreuv, I. 1913. Relation between size of host and sex in Ichneumonidae. *C. R. Soc. Biol.*, 74: 695-699.

Chundurwar R.D. 1975. Fecundity and life table studies on *Eriborus trochanteratus* Morley (Hymenoptera : Ichneumonidae) *Jour. Indian Potato Associ.*, 2(1) : 42-45.

Church, N.S. 1955. Hormones and diapause in *Cephus*. *Can. J. Zool.*, 33, 339-369.

Doutt, R.L. 1959. The biology of parasitic Hymenoptera. *A Rew. Ent.*, 4, 161-182.

Fisher, R.C. 1959. Life history and ecology of *Horogenes chrysostictes* Gmelin (Hymenoptera : Ichneumonidae), a parasite of *Ephestia sericurium* Scott. (Lepidoptera : Phycitidae). *Can. J. Zool.*, 37(1), 429-446.

Flanders, S.E. 1942. Diapause in parasitic insects. *J.Econ. Ent.*, 35, 607.

Gangrade, G.A. 1964. On the biology of *Campoletis perdistinctus* (Hymenoptera : Ichneumonidae) in Madhya Pradesh. *Ann. ent. Soc. Am.*, 57, 570-574.

Gupta S. and Gupta V.K. 1983. Ichneumonologia orientalis, Part-IX, The tribe Gabuniini (Hymenoptera : Ichneumonidae). *Oriental Ins. Monogr.*, 10, 1-313.

Gupta, V.K. 1987. Catalogue of the Indo-Australian Ichneumonidae. *Mem. Amer. Ent. Inst.*, 41, Part I and II, 1-1210.

Leius, K. 1961. Influence of food on fecundity and longevity of adults of *Itoplectis conquisitor* (Say) (Hymenoptera : Ichneumonidae). *Can. Ent.*, 93, 771-780.

Leong, G.K.L. and E.R. Oatman 1968. The biology of *Campoplex haywardii* (Hymenoptera : Ichneumonidae), a primary parasite of the potato tuber worm. *Ann. ent. Soc. Am.*, 61(1), 26-36.

Lingren, P.D., Guerra, R.J., Nickelson, J.W. and C. White 1970. Hosts and host age preference of *Campoletis perdistinctus*. *J. Econ. Ent.*, 63, 518-522.

Lingren, P.D., Wilson F.D. and R.I. Sailer 1977. Cross mating of *Campoletis sonorensis* and *Campoletis chlorideae*. *Ann. ent. Soc. Am.*, 70(4), 442-446.

Marchal P. 1936. Diapause in host and parasite : *Trichogramma. Ann. Epiphyt. Phytogen.*, 2, 447-550.

Mashennikova, V. 1958. Diapause in *Apanteles* influenced by the host. *A. Rew. Ent.* (URSS), 37, 538-545.

Morley C. 1913. Fauna of British India including Ceylon and Burma. Hymenoptera : Ichneumonidae. 3(1), 1-531.

Nikam P.K. and C.D. Basarkar 1978. Studies on the effect of temperature on the development of *Campoletis chlorideae* Uchida (Ichneumonidae), an internal larval parasitoid of *Heliothis armigera* (Hubn.). *Entomon.,* 3(2), 307-308.

Nikam P.K. and G.K. Gosawi 1982. Daily oviposition rate, fecundity and longevity of *Eriborus argentiopilosus* (Cameron) (Hymenoptera : Ichneumonidae) a larval parasitoid of *Heliothis armigera* (Hubn.) *Indian J. Parasitol.,* 6(2), 259-262.

Nikam, P.K. 1980. Studies on Indian species of *Enicospilus stephens* (Hym. :Ichneumonidae) *Oriental Ins.,* 14(2), 131-219.

Oatman, E.R. and G.R. Platner 1974. The biology of *Temelucha* sp. *Platensis* Group (Hymenoptera : Ichneumonidae), a primary parasite of the potato tuber worm. *Ann. ent. Soc. Am.,* 67, 275-280.

Parkar, H.L. 1935. *U.S.D.A., Tech. Bull No. 477,* 17 pp.

Prebble, M. L. 1941. Diapause : *Gilpinia,* Hymenoptera. *Canad. J. Res.,* 19, 295-454.

Quednau, T.W. and Guverment, H. 1975. Observations on mating and oviposition behaviour in *Priopoda nigricollis* (Hym; Ichneumonidae), a parasite of the birch leafminer *Fenusa pusilla* (Hym; Tenthredinidae). *Can. Ent.* 107, 1199-1204.

Rojas - Rousse and Benoit, M. 1977. Morphology and biometry of larval instars of *Pimpla instigator* (F.) (Hymenoptera : Ichneumonidae), *Bull. Ent. Res.,* 67, 129-141.

Salt, G. 1941. Effect of hosts on parasites. *Biol. Rev.,* 16, 239-264.

Sathe T.V. 1988. Biology of *Diadegma trichoptilus* (Cameron) (Hym. : Ichneumonidae), a larval parasitoid of *Exelastis atomosa* Walsingham. *J. Curr. Biosci.,* 5, 37-40.

Sathe T.V. 1990. Biology of *Diadegma argenteopilosa* (Cameron) (Hym. : Ichneumonidae), an internal larval parasitoid of *Spodoptera litura* (Fab.) *The Entomologist,* 109, 2-7.

Sathe T. V. 2004. Vermiculture and organic farming. Daya Publishing House, New Delhi. 1-122 pp.

Sathe, T.V. and A.M. Bhosale 2011. *Plytella xyllostella* (Lihn.) density requirement for maximum progeny production of *Diadegma insularae* Cameron (Hym.; Ichneumonidae). *Asian J. Anim. Sci.*, 6(2), 212-214.

Sathe, T.V. and Margaj G.S. 2001. Cotton pests and biocontrol agents. Daya Publi. House, Delhi. 1-157 pp.

Sathe T. V., S. A. Inamdar and R. K. Dawale 2003. Indian pest parasitoids. Daya Publishing House, New Delhi. 1-145 pp.

Schmidt G.T. 1975. Host acceptance behaviour of *Campoletis sonorensis* towards *Heliothis rea. Ann. ent. Soc. Am.*, 67, 835-844.

Schneiderman H.A. and J. Horowitz 1958. Diapause in *Mormoniella* and *Tritneptis. J. Expt. Biol.*, 35, 520-551.

Short, J.R.T. 1970. On the classification of the final instar larvae of Ichneumonidae (Hymenoptera) supplement. *Trans. R. ent. Soc. Lond.*, 122, 185-210.

Simmonds F.J. 1946. Diapause in Hymenopterous parasites. *Bull. Ent. Res.*, 37, 95-97.

Thompson W.R. 1945. A catalogue of the parasites and predators of insect pests (section-I) 6 : 25.

Tikar D.T. and K.R.Thakare 1961. Bionomics, biology and immature stages of an Ichneumonid *Horogenes fenestralis* Holmgren, a parasite of common caterpillar. *Indian J. Ent.,* 23 (11), 116-124.

Townes W. (1970). The genera of Ichneumonidae. Part-II. *Mem. Amer. Ento. Inst., 12,* 1-537.

Townes, H., Townes, M. and V.K. Gupta (1961). A catalogue and classification of Indo-Australian Ichneumonidae. *Mem. Amer. Ento. Inst.,* 1, 1-522.

Yeargan K.V. and Latheef M.A. 1976. Host parasitoid density relationship between *Hypera postica* (Coleop., Curculionidae) and *Bothyplectes curculionis* (Hym. : Ichneumonidae). *J. Kanas. Cant.* 49, 551-556.

Index